U0489069

河北自然资源图鉴

Hebei Natural Resources Illustrations Handbook

章洪涛　主编
Zhang Hongtao　Editor-in-Chief

中国摄影出版传媒有限责任公司
China Photographic Publishing & Media Co., Ltd.
中国摄影出版社

人与自然和谐共生

Harmonious Coexistence Between Man and Nature

编委会

主　编：章洪涛

副主编：杨淑梅　任白羽

编写者：贾　彬　姜慧婕　孙玉霞　戴祥祥　郑新龙
　　　　梁小珍　苗晓昕　郭天翔　宋海娟　朱鹏涛
　　　　陈晓甫　陈晓雨　韩　阳

EDITORIAL BOARD

Editor-in-Chief: Zhang Hongtao

Deputy Editor: Yang Shumei, Ren Baiyu

Written by: Jia Bin, Jiang Huijie, Sun Yuxia, Dai Xiangxiang, Zheng Xinlong
Liang Xiaozhen, Miao Xiaoxin, Guo Tianxiang, Song Haijuan
Zhu Pengtao, Chen Xiaofu, Chen Xiaoyu, Han Yang

绿水青山就是金山银山

Green mountains and clear water are equal to mountains of gold and silver

目录 CONTENTS

前　言	Foreword	28
国土空间规划	Territorial Spatial Planning	30
土地资源	Land Resources	46
矿产资源	Mineral Resources	70
森林资源	Forest Resources	80
草原资源	Grassland Resources	94
湿地资源	Wetland Resources	104
水资源	Water Resources	116
海洋资源	Marine Resources	126
野生动物资源	Wildlife Resources	148
自然保护区资源	Nature Reserve Resources	174
国家公园资源	National Park Resources	200
美景赏析	Appreciation of Beautiful Scenery	214
后　记	Postscript	230

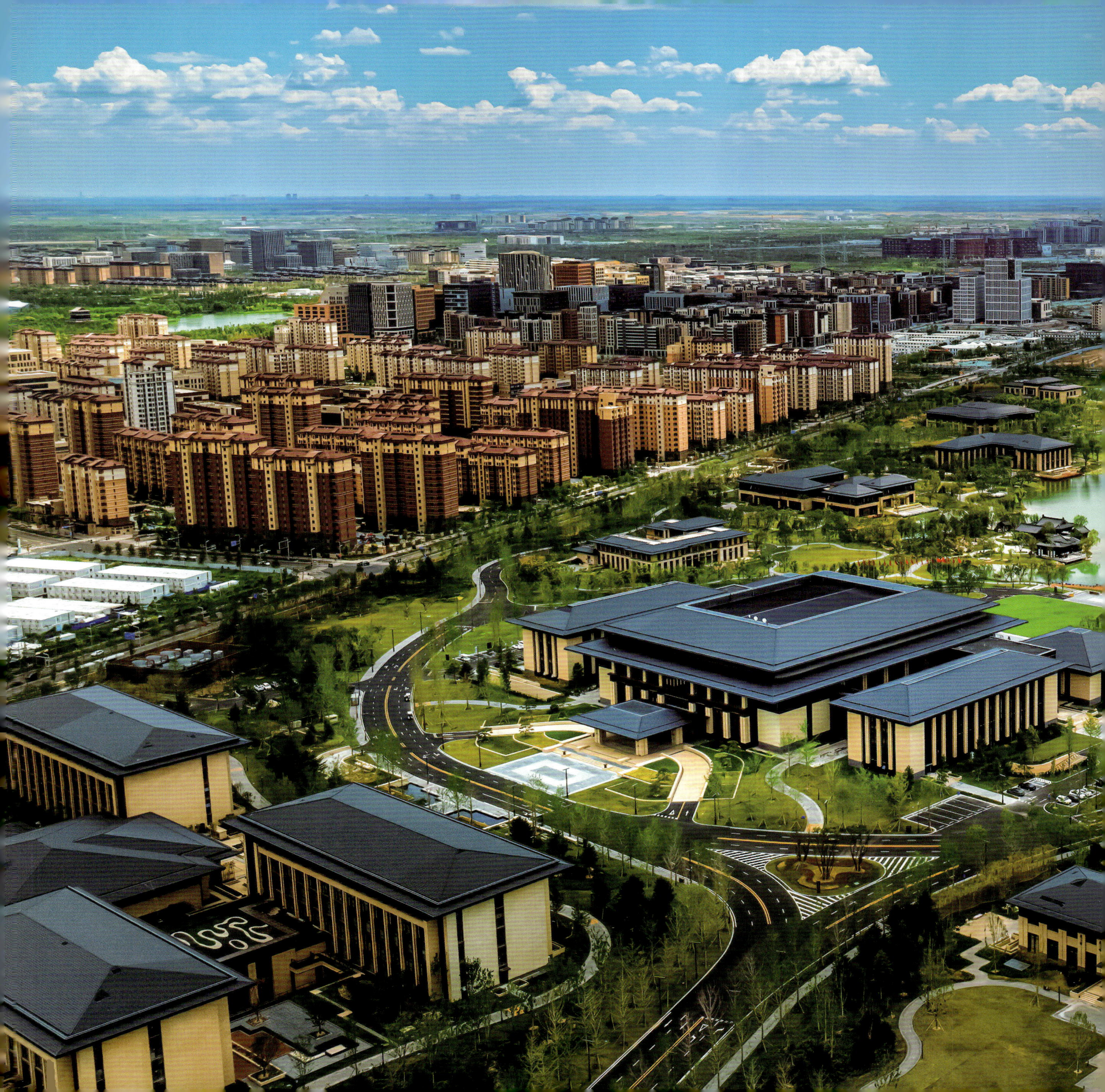

国土空间规划

高标准高质量建设雄安新区，有效承接北京非首都功能疏解，着力建设具有核心竞争力的产业集群。

《雄安·未来已来》，雄安新区容东安置区，摄影：薛志勇

Territorial Spatial Planning
Build the Xiong'an New Area to high standards and quality, effectively undertake the relocation of non-capital functions from Beijing, and focus on building industrial clusters with core competitiveness.
Xiong'an--The Future Is Already Here, Rongdong esettlement Area in Xiong'anNew Area
Photographed by Xue Zhiyong

■ 土地资源

河北约有 3866.67 平方千米盐碱地，其中盐碱耕地 3800 平方千米，是我国盐碱地面积较大的省份之一。
《夏收》，黄骅市旱碱麦田，摄影：郑勇

Land Resources
Hebei has about 3866.67 square kilometers of saline-alkali land, of which more than 3800 square kilometers million mu are saline-alkali farmlands, ranking it one of the provinces with the largest area of saline-alkali land in China.
Summer Harvest, Saline-alkali Wheat Field in Huanghua City
Photographed by Zheng Yong

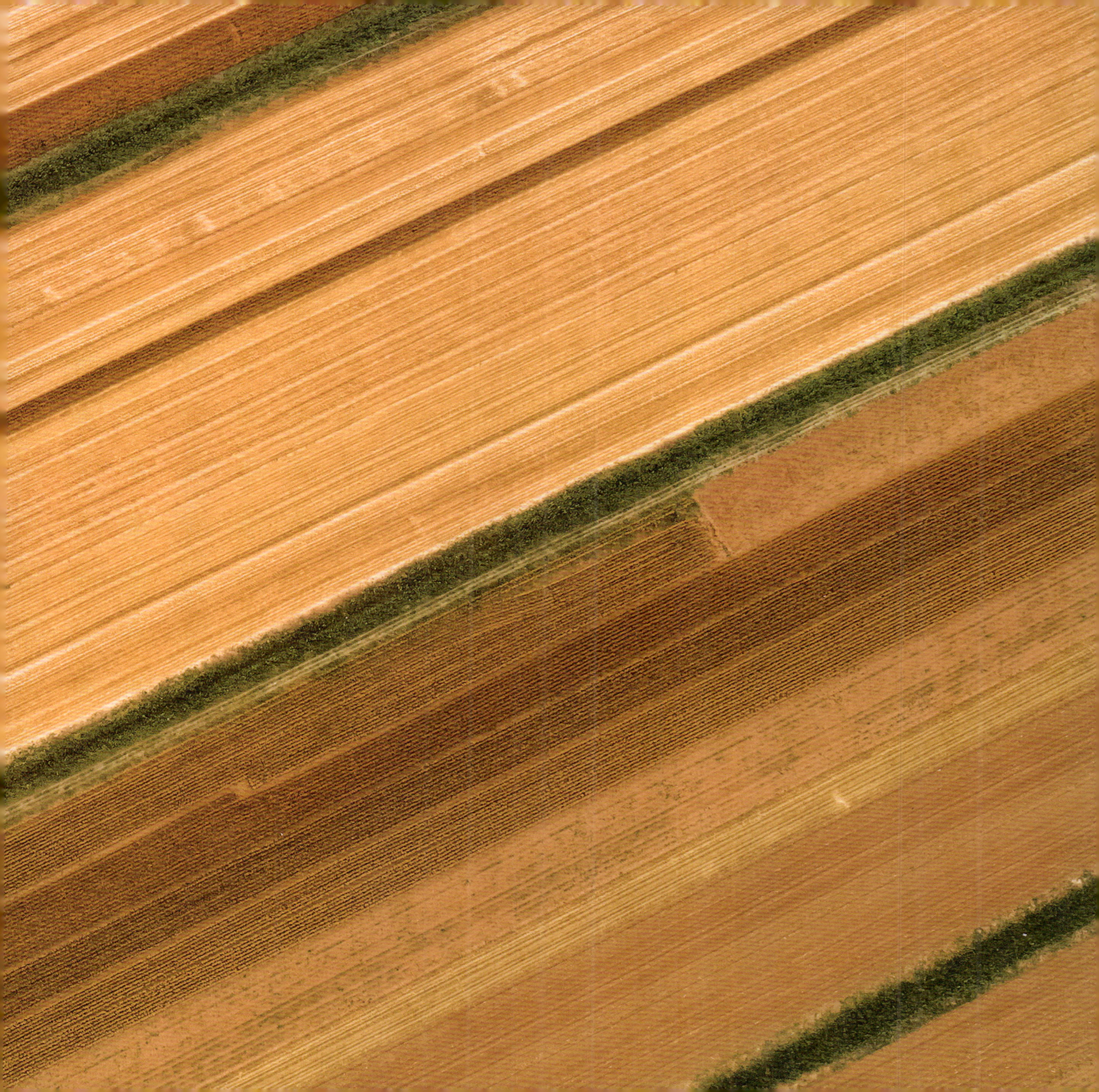

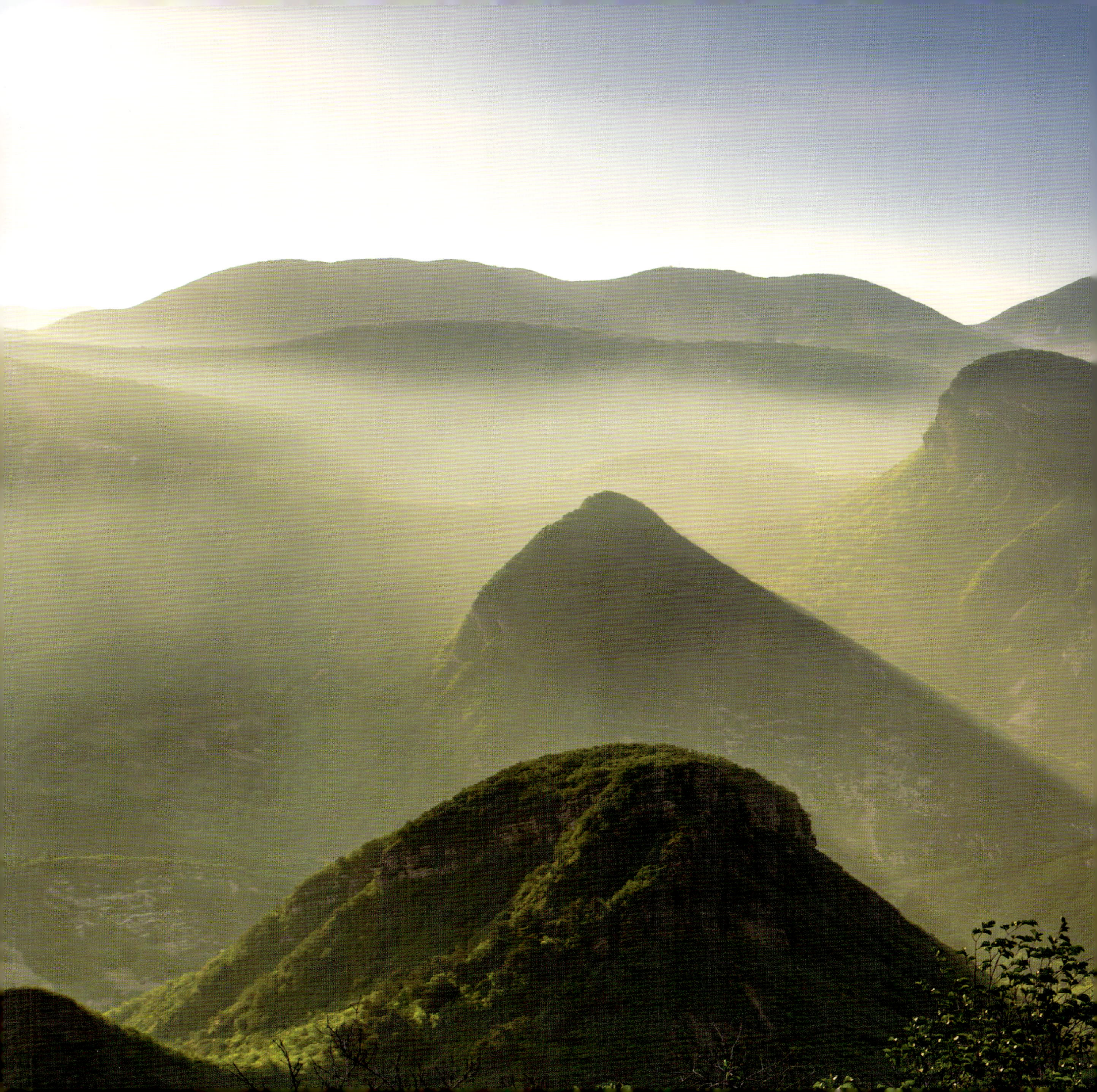

■ 矿产资源

　　三河市东北部山区经过矿山修复治理，成为该市绿色发展的新引擎。
《叠峦》，三河市东北部山区，摄影：李连明

Mineral Resources
The mountainous region in the northeastern part of Sanhe City has emerged as a new engine for the city's green development after undergoing mine restoration and management.
The Stacked Hills, Mountainous Area in Northeast Sanhe City
Photographed by Li Lianming

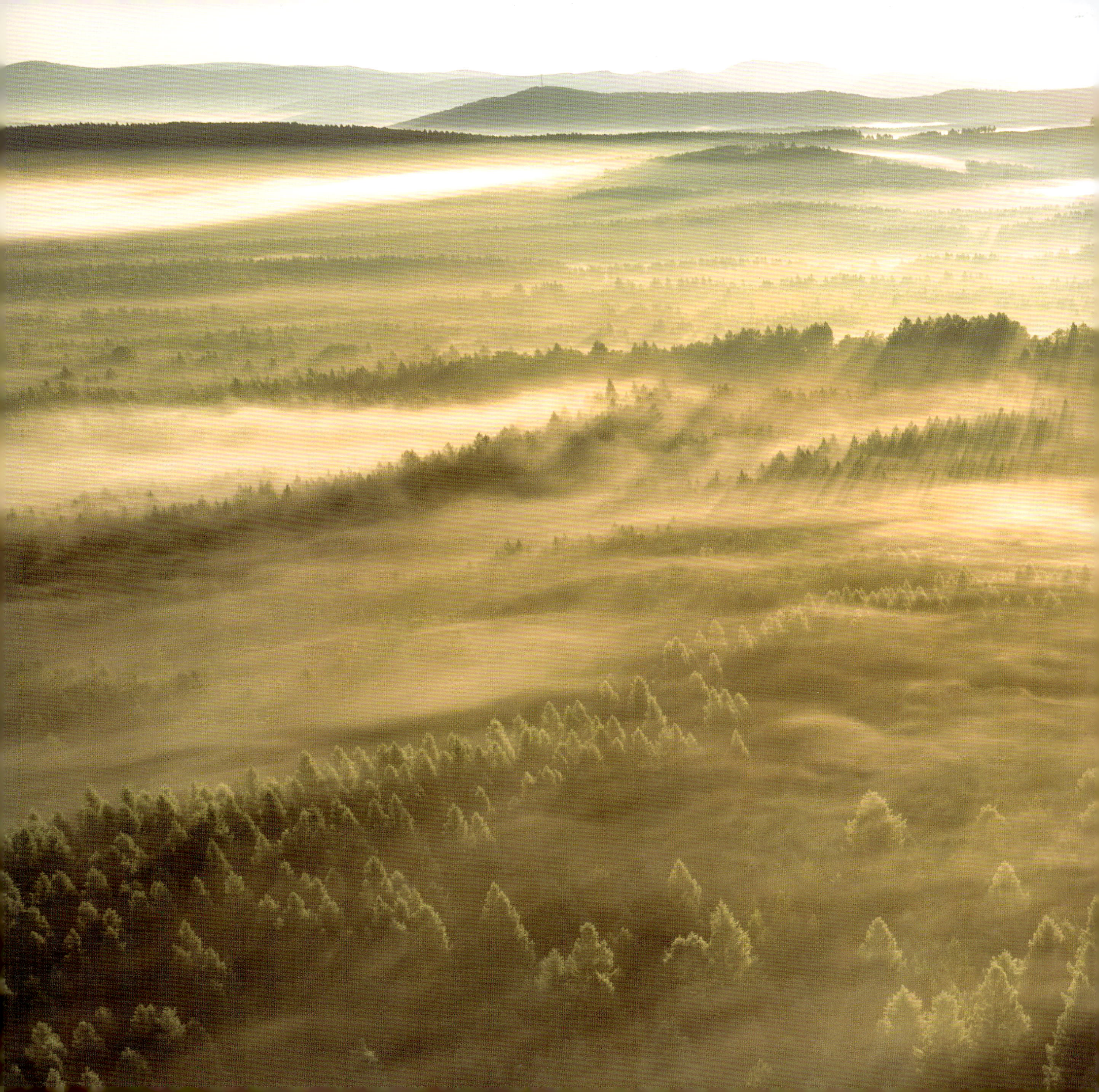

森林资源

塞罕坝机械林场总经营面积约为933.33平方千米，其中林地面积约767.33平方千米，林木蓄积量约有1036.8万立方米，森林覆盖率达82%，成为京津地区的"水源卫士"、生态屏障。

《日出塞罕坝》，承德市塞罕坝机械林场，摄影：林树国

Forest Resources

Saihanba Mechanical Forest Farm covers a total management area about 933.33 square kilometers, of which approximately 767.33 square kilometersare forest land. The volume of timber stock is about about 10.368 million cubic meters, and the forest coverage rate is 82%, earning it the title of the Water Source Guardian and serving as an ecological barrier for the Beijing-Tianjin region.

Sunrise at Saihanba, Saihanba Machinery Forest Farm in Chengde City
Photographed by Lin Shuguo

草原资源

　　河北省是我国北方的重要草原区，草原是其绿色优势、生态屏障、资源禀赋，全省草地面积约有 19472.67 平方千米。
《坝上草原》，承德市，摄影：佚名

Grassland Resources
Hebei Province is an important grassland area in northern China, where the grassland is its green advantage, ecological barrier, and resource endowment. The total area of grassland in the province is about 19472.67 square kilometers.
Grassland on the Dam, Chengde City
Photographed by Anonymous

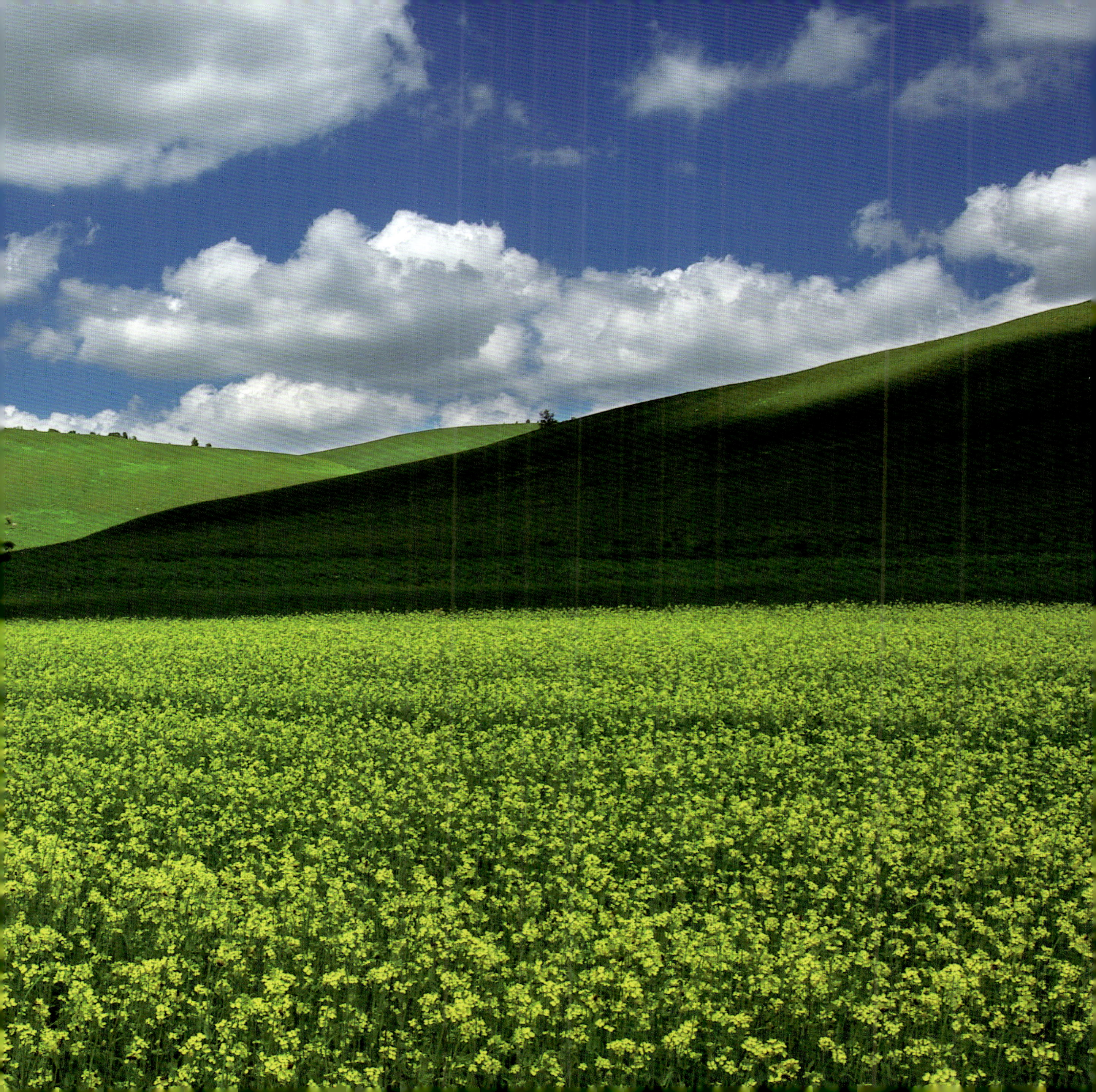

湿地资源

　　闪电河国家湿地公园的动植物资源丰富，有植物 297 种，陆生野生脊椎动物 274 种，占全省陆生脊椎动物总量的 41.2%。其中，鸟类多达 222 种，较 2019 年增加 35 种。

《闪电河国家湿地公园》，张家口市沽源县城东部，摄影：赵永春

Wetland resources
The Lightning River National Wetland Park is rich in animal and plant resources, with 297 species of plants and 274 species of terrestrial wild vertebrates, accounting for 41.2% of the total terrestrial vertebrates in the province. Among them, there are as many as 222 species of birds, an increase of 35 species from 2019.
Lightning River National Wetland Park, eastern part of Guyuan County, Zhangjiakou City
Photographed by Zhao Yongchun

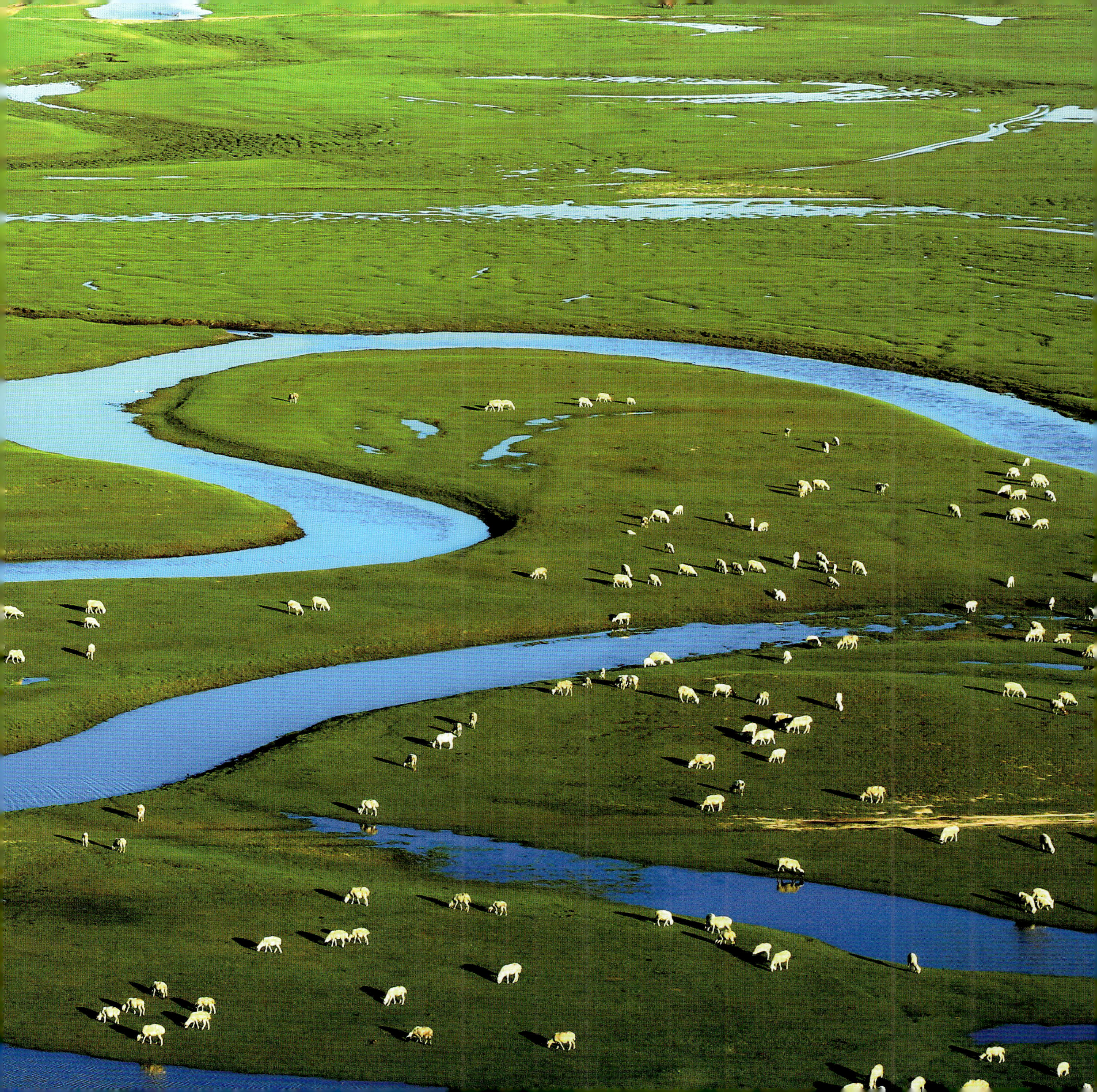

■ 水资源

秦皇岛燕塞湖，因地处燕山要塞而得名。这里位于山海关城西面约3.5公里处，生态环境优美典雅，是长寿山国家森林公园和秦皇岛柳江国家地质公园的重要组成部分。
《秦皇岛燕塞湖》，秦皇岛市山海关区城西，摄影：袁晓虹

Water resources
Qinhuangdao Yansai Lake, named for the Yanshan Fortress, is situated about 3.5 kilometers west of Shanhaiguan City. It has a beautiful and elegant ecological environment and is an important part of Changshoushan National Forest Park and Qinhuangdao Liujiang National Geopark.
Qinhuangdao Yansai Lake, West of Shanhaiguan District, Qinhuangdao City
Photographed by Yuan Xiaohong

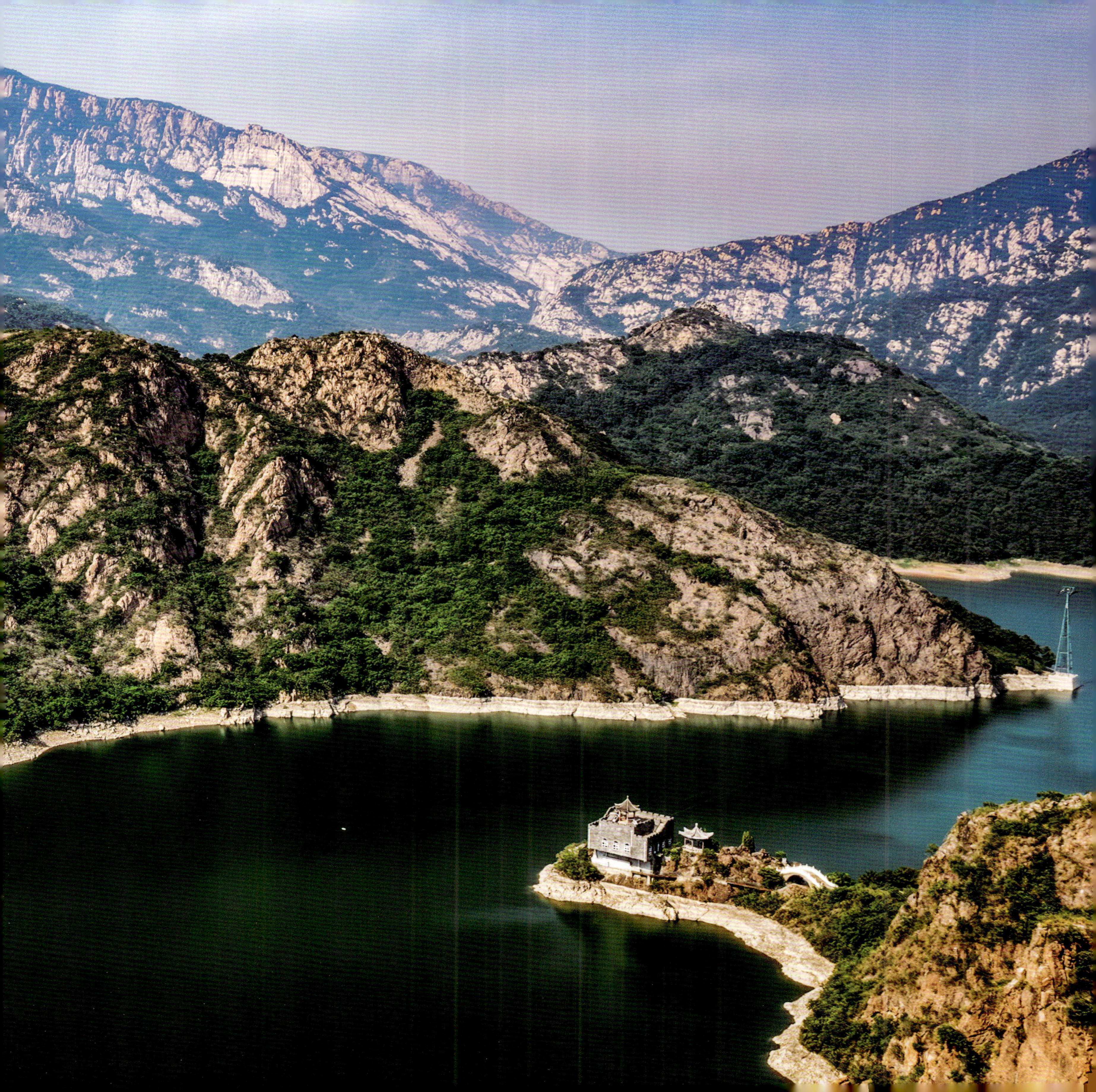

海洋资源

河北省海洋经济资源类型多样，海岸线开发程度较高，主要利用类型有渔业、盐业、交通运输、工业、旅游、矿产与能源。近年来，海上光伏、海上风电方面有新的发展和突破。

《耕海》，秦皇岛市北戴河新区洋河口，摄影：崔重辉

Marine Resources

Hebei Province boasts a diverse range of marine economic resources and a highly developed coastline. The main types of utilization include fisheries, salt production, transportation, industry, tourism, minerals and energy. In recent years, there have been new developments and breakthroughs in offshore photovoltaics and wind power.

Marine Cultivation, Yanghekou, Beidaihe New District, Qinhuangdao City
Photographed by Cui Chonghui

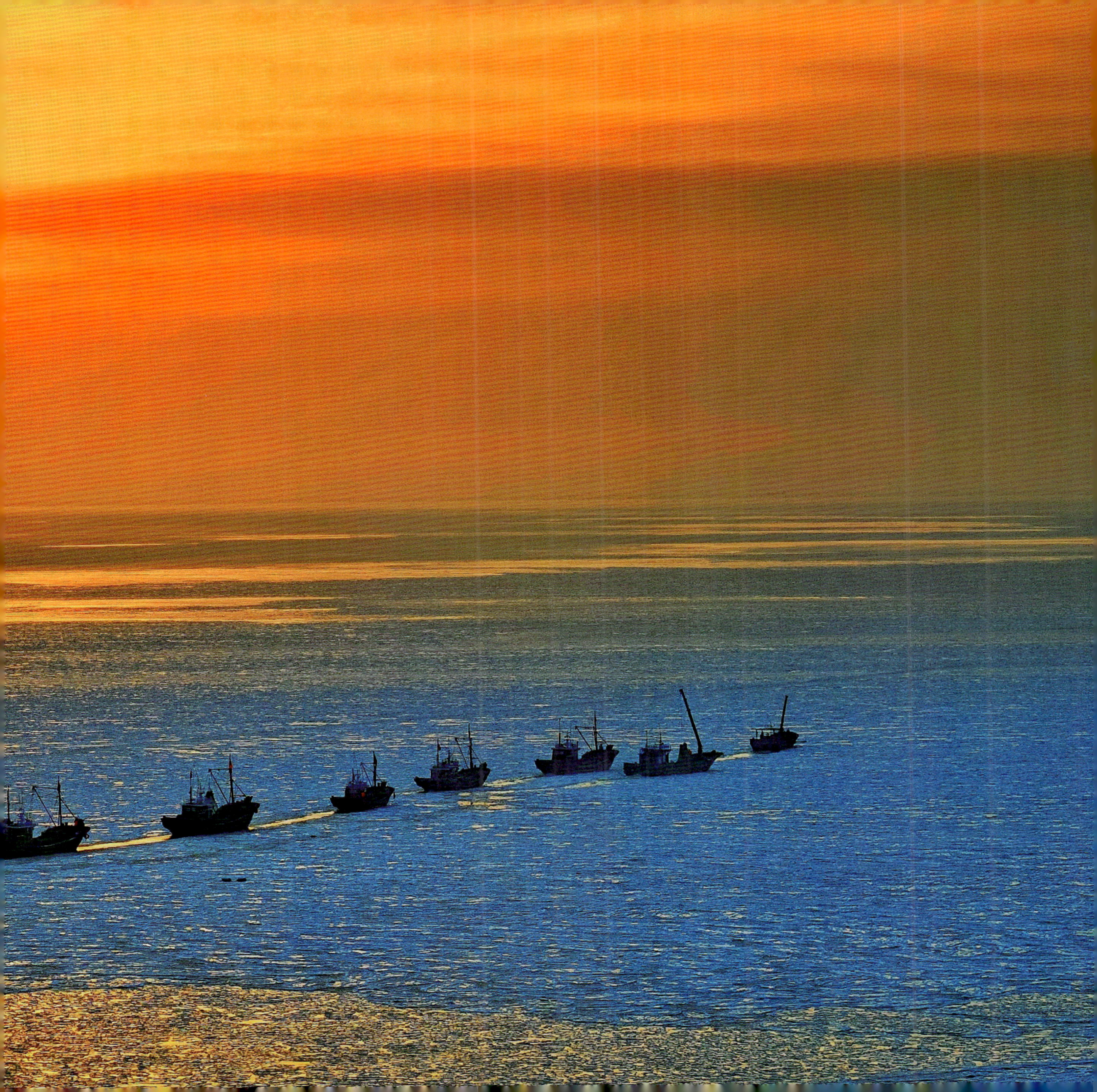

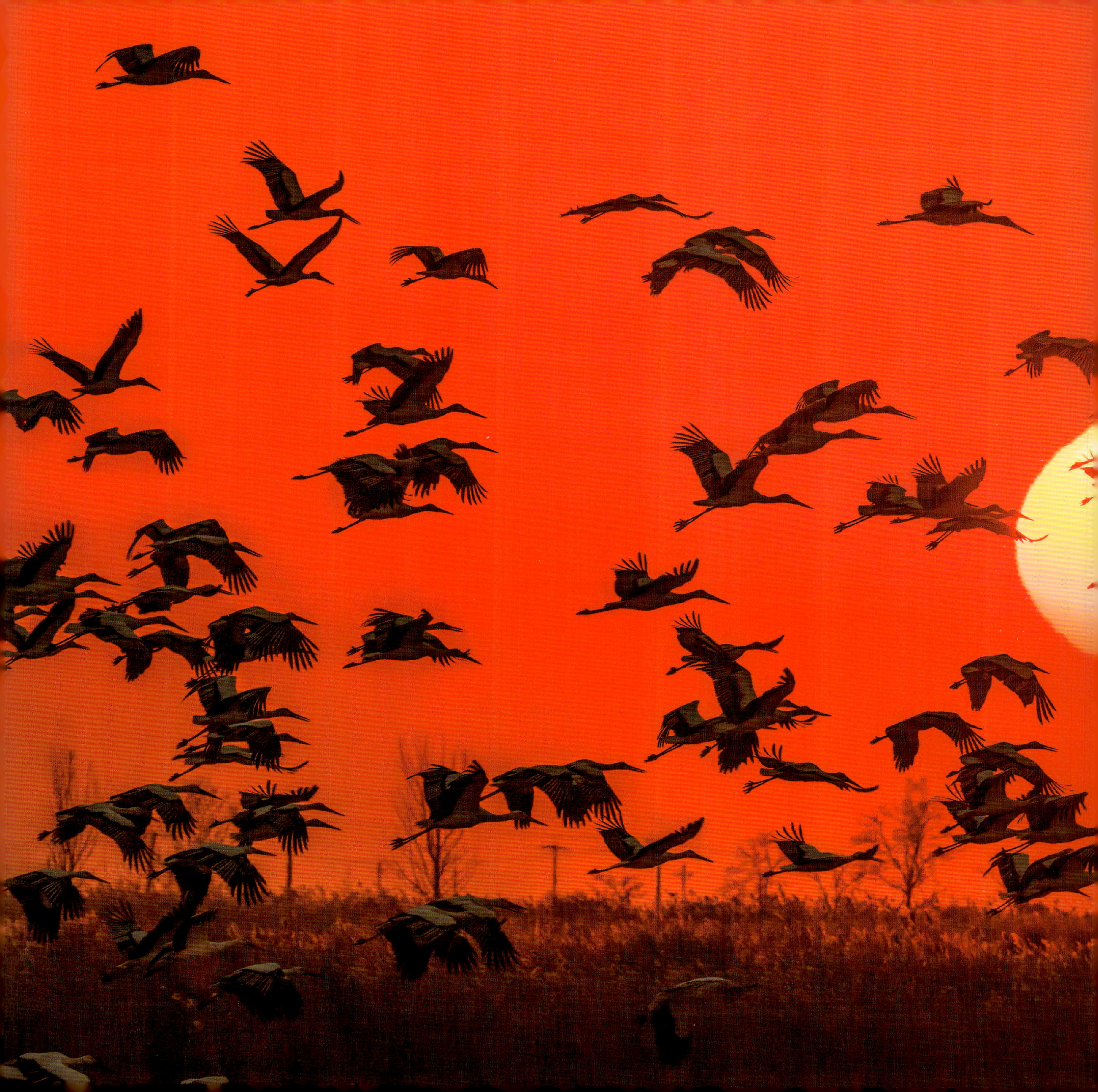

■ 野生动植物资源

河北省现有陆生野生脊椎动物 605 种，包括哺乳类 87 种，鸟类 486 种，两栖类 8 种，爬行类 24 种。画面中捕捉到的是一群东方白鹳飞过曹妃甸湿地。
《曹妃甸湿地》，唐山市曹妃甸区，摄影：张玉成

Wild Fauna and Flora Resources
Hebei Province currently has 605 species of terrestrial wild vertebrates, including 87 species of mammals, 486 species of birds, 8 species of amphibians, and 24 species of reptiles. The scene captures a group of Oriental White Storks flying over the Caofeidian Wetlands.
Caofeidian Wetlands, Caofeidian District, Tangshan City
Photographed by Zhang Yucheng

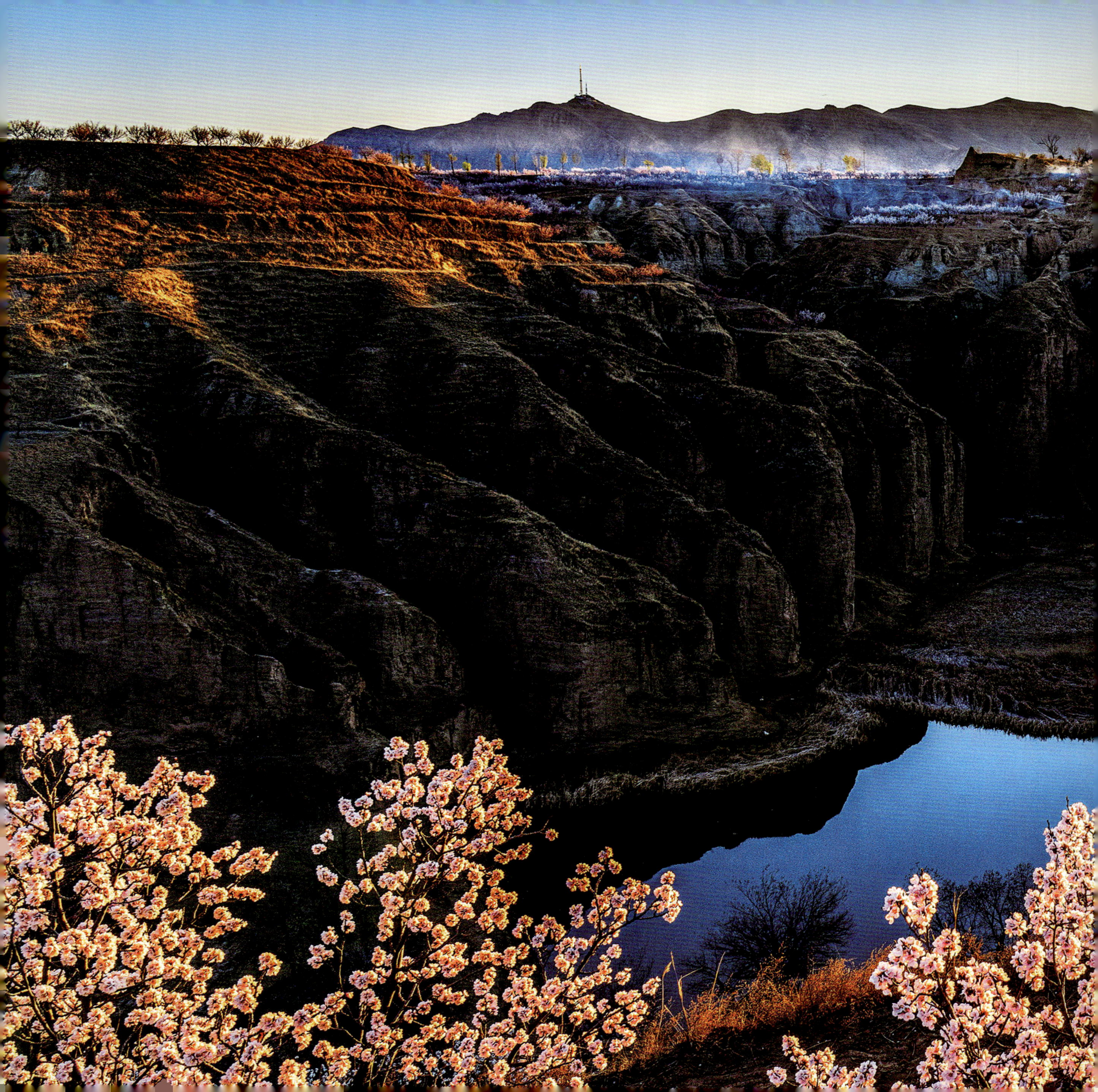

自然保护区资源

　　河北泥河湾国家级自然保护区位于泥河湾盆地的东部，主要保护对象为新生代典型地层剖面、晚新生代地层中的哺乳动物化石及其主要发掘遗址。地层中的人类文化遗迹和古人类活动遗址、晚新生代地层中的软体和微体动物化石、湖相叠层石、发育在泥河湾地层中的特殊地貌景观，属地质遗迹类型自然保护区。

《泥河湾国家级自然保护区》，张家口市阳原县，摄影：潘辉峰

Nature Reserve Resources

Nihewan National Nature Reserve in Hebei Province is located in the eastern part of the Nihewan Basin. Its primary conservation focus is on typical Cenozoic stratigraphic profiles, mammal fossils from the late Cenozoic strata and their main excavation sites, human cultural relics and ancient human activity sites within the strata, soft-bodied and microfossil invertebrates from the late Cenozoic strata, lacustrine stromatolites, and unique geomorphological landscapes developed in the Nihewan strata. It is a nature reserve of geological heritage type.

Nihewan National Nature Reserve, Yangyuan County, Zhangjiakou City
Photographed by Pan Huifeng

国家地质公园资源

赞皇嶂石岩国家地质公园位于河北省石家庄市赞皇县，公园内最为典型的地质景观特征是嶂石岩地貌和元古宇长城系砂岩中的层理与层面构造（交错层构造、波纹构造）。

《赤壁丹崖九女峰》，石家庄市赞皇县嶂石岩村，摄影：孙利人

National Geopark Resources
Zanhuang Zhangshiyan National Geopark in Hebei Province is located in Zanhuang County, Shijiazhuang City. The most typical features of the park are the Zhangshiyan landform and the bedding and layer structures (including cross-bedding structure, ripple structures) in the sandstones of the Proterozoic Changcheng System.
Red Cliff Nine Women Peak, Zhangshiyan VillageZanhuang County, Shijiazhuang City
Photographed by Sun Liren

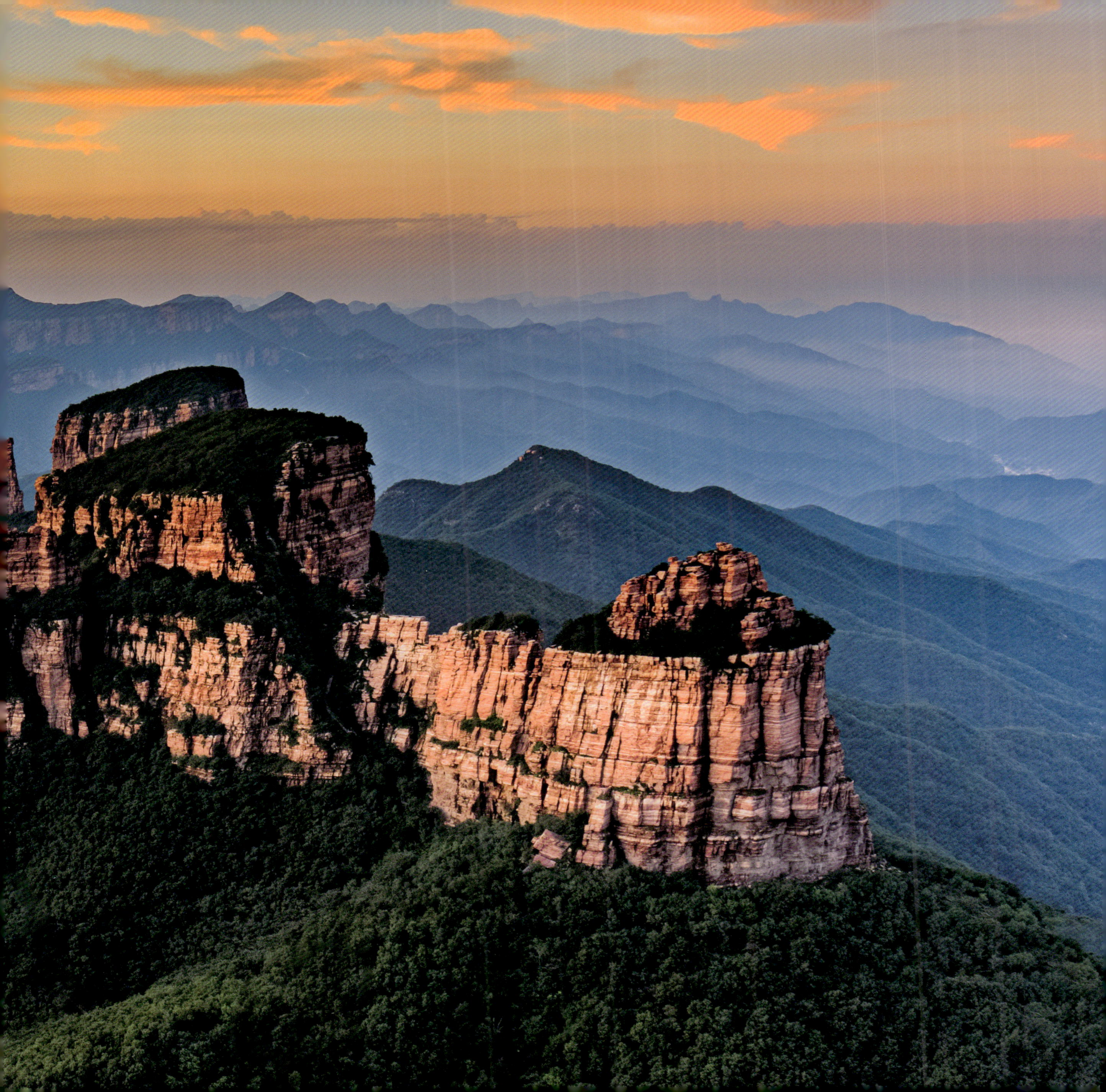

前　言

河北，简称"冀"，别称"燕赵"，曾用名"直隶"。篆书"冀"的整个字形就像一只展翅飞翔的大鸟，努力向上，充满希望。

河北省自然资源丰富，高原、山地、丘陵、盆地、平原、海滨等样样齐备。太行绵延，燕山逶迤，两大山脉以"人"字状交汇，撑起京、津、冀三地的生态屏障。平原辽阔坦荡，沃野千里，一季耕作，一场丰收，粮蔬瓜果循季而至，丰富着人们的日常生活。湖泊河流，海浪洪波，八方汇聚，接续着数代人海晏河清的美好愿望。森林草原，沙滩湿地，四季更迭，为河北大地增添了别样风情。日出云散，风吹花开，镜头定格，大自然赋予的每一个美妙瞬间，无不呈现出人与自然和谐共生的生动画卷。勤劳勇敢的河北人民在这片土地上传承发展着热爱自然资源，珍惜每一寸土地，合理利用自然资源的理念日益深入人心。

习近平总书记在视察河北塞罕坝林场时指出，全党全社会要坚持绿色发展理念，弘扬塞罕坝精神，持之以恒推进生态文明建设，一代接着一代干，驰而不息，久久为功，努力形成人与自然和谐发展新格局，把伟大祖国建设得更加美丽，为子孙后代留下天更蓝、山更绿、水更清的优美环境。为深入贯彻习近平生态文明思想和习近平总书记关于自然资源管理的重要论述，全面展示河北自然资源特色风貌和管理成效，探索构建全省自然资源领域数据融合、信息互通的新途径，河北省自然资源厅立足部门职责，组织力量编撰出版了这部《河北自然资源图鉴》，旨在引导全社会坚持"山水林田湖草生命共同体"理念，全面贯彻节约优先、保护优先、自然恢复为主的方针，进一步加强自然资源保护利用，稳步推进生态文明建设，对支撑河北省自然资源管理、服务经济社会发展具有重要意义。

《河北自然资源图鉴》包括"国土空间规划""土地资源""矿产资源""森林资源""草原资源""湿地资源""海洋资源""野生动物资源""自然保护区资源""国家公园资源""水资源""美景赏析"12个专题，采用河北省自然资源厅、河北省林业和草原局等单位和部门的基础资料，收录专业摄影作品和近年来河北省自然资源摄影部分优秀作品，精选彩色图片205幅，各类图表30余张，综合性文字2.6万字，中英文对照，图文并茂，系统呈现了河北自然资源空间分布、开发利用、保护监管和与之相关的丰富信息。这部生动解读河北自然资源的图书，具有很强的可读性、观赏性、实用性，是认识河北、了解河北的绝佳窗口。

驻足回眸，你会发现，山河相望，岁月静好，河北正在建设人与自然和谐共生的现代化新长征道路上昂扬奋进！大河之北这片纯朴厚重的土地日新月异，希望无限！

由于编者水平有限，《河北自然资源图鉴》难免存在疏漏和不足，敬请各位读者批评指正，以期进一步修改完善！

河北省自然资源厅宣传中心

Foreword

Hebei, referred to as Ji, also known as Yan Zhao, used to be known as Zhili. The glyph of the seal script "Ji" is like a big bird flying with wings, striving upwards and full of hope.

Hebei Province is rich in natural resources, including plateaus, mountains, hills, basins, plains and seashores. Taihang stretches, Yanshan Mountain is long, and the two mountain ranges converge in the shape of chinese"person", supporting the ecological barrier of Beijing, Tianjin and Hebei. The plains are vast and open, fertile fields for thousands of miles, a season of farming, a bumper harvest, grain, vegetables, melons and fruits come according to the seasons, enriching people's daily life. Lakes and rivers, waves and waves, gathering in all directions, continue the good wishes of several generations peace and prosperity. Forests and grasslands, beaches and wetlands, the change of seasons, add a different style to the land of Hebei. The sun rises and the clouds disperse, the wind blows and the flowers bloom, the camera freezes, and every wonderful moment given by nature presents a vivid animation of the harmonious coexistence of man and nature. The industrious and brave people of Hebei have inherited and developed the concept of loving natural resources, cherishing every inch of land, and rationally using natural resources in this land.

When General Secretary Xi Jinping inspected the Saihanba Forest Farm in Hebei Province, he pointed out that the whole party and the whole society should adhere to the concept of green development, carry forward the spirit of Saihanba, and persistently promote the construction of ecological civilization. In order to thoroughly implement Xi Jinping Thought on Ecological Civilization and General Secretary Xi Jinping's important exposition on natural resource management, comprehensively display the characteristics and management effectiveness of Hebei's natural resources, and explore new ways to build data integration and information exchange in the field of natural resources in the province, the Department of Natural Resources of Hebei Province has organized forces to compile and publish this *Hebei Natural Resources Illustrations Handbook* based on the responsibilities of the department, aiming to guide the whole society to adhere to the concept of "life community of mountains, rivers, forests, fields, lakes and grasses", and fully implement the principle of giving priority to conservation, protection and natural restoration. It is of great significance to further strengthen the protection and utilization of natural resources and steadily promote the construction of ecological civilization to support the management of natural resources and serve the economic and social development of Hebei Province.

Hebei Natural Resources Illustrations Handbook includes 12 topics: "Territorial Spatial Planning" "land resources" "mineral resources" "forest resources" "grassland resources" "wetland resources" "marine resources" "wildlife resources" "nature reserve resources" "national park resources" "water resources" and "Appreciation of Beautiful Scenery", using the basic data of Hebei Provincial Department of Natural Resources, Hebei Provincial Forestry and Grassland Bureau and other units and departments, including professional photography works and some excellent works of Hebei Natural Resources Photography Exhibition in recent years, 205 selected color pictures, more than 30 charts of various kinds, 26,000 words of comprehensive text, Chinese and English, combining pictures and text, systematically presenting the spatial distribution, development and utilization, protection and supervision of Hebei's natural resources and rich information related to it. Vividly interpreting Hebei's natural resources, this book is available in both hardcover and litecover versions, with strong readability, ornamentality, practicability, is an excellent window to know Hebei and understand Hebei.

Stop and look back, you will find that the mountains and rivers are facing each other, the years are quiet, Hebei is building a modern new Long March road of harmonious coexistence between man and nature, and the simple and heavy land in the north of the river is changing with each passing day, and the hope is unlimited!

Due to the limited level of editors, there are inevitably omissions and deficiencies in the *Hebei Natural Resources Illustrations Handbook*, so please criticize and correct readers in order to further revise and improve.

Publicity Center of Hebei Provincial Department of Natural Resources

国土空间规划

国土空间规划是国家空间发展的指南、可持续发展的空间蓝图，是各类开发保护建设活动的基本依据。2023年12月9日，国务院批复《河北省国土空间规划（2021—2035年）》（以下简称《规划》）。

河北省地处京畿要地，是实施京津冀协同发展战略的关键区域之一。河北省坚决贯彻落实京津冀协同发展战略，牢牢牵住疏解北京非首都功能"牛鼻子"，共同建设以首都为核心的世界一流城市群。

落实京津冀协同发展总体格局要求。与京津共同推进现代化首都都市圈建设；支持环京津市县与北京、天津联动发展；强化京津、京保石、京唐秦三条发展轴带动作用；促进环京津核心功能区、沿海率先发展区、冀中南功能拓展区、冀西北生态涵养区四大功能区协调联动；培育壮大石家庄、唐山、保定、邯郸等区域中心城市。

高标准、高质量建设雄安新区。面向重大国家战略需求，有效承接北京非首都功能疏解，着力建设具有核心竞争力的产业集群；支持雄安新区优势产业向周边地区拓展，形成产业集群；推动雄安新区及周边交通基础设施互联互通，强化协同发展基础保障。

打造后冬奥时代绿色高质量发展新高地。支持张北地区发展后奥运经济，加强场馆赛后综合利用，培育形成具有全球影响力的国际冰雪运动与休闲旅游胜地。

筑牢安全发展的空间基础。到2035年，河北省耕地保有量不低于58673.33平方千米；生态保护红线面积不低于3.639万平方千米，其中海洋生态保护红线面积不低于0.11万平方千米；城镇开发边界扩展倍数控制在基于2020年城镇建设用地规模的1.3倍以内；单位国内生产总值建设用地使用面积下降不少于41%；大陆自然岸线保有率不低于国家下达的任务；用水总量不超过国家下达的指标；除国家重大项目外，全面禁止围海、填海；严格无居民海岛的管理。

统筹优化农业、生态、城镇和海洋等功能空间布局。在农业空间安排上，构建以太行山山前平原、燕山山前平原和黑龙港低平原地区等农产品集中产区为主体的农业生产格局，夯实华北粮仓空间基础。

在生态空间安排上，提升燕山—太行山区、京津冀北部—辽河源、渤海海岸带等重要地区生态功能，持续抓好北方防沙带等生态保护和修复重点工程建设。

在城镇空间安排上，围绕建设京津冀世界一流城市群，强化石家庄都市圈引领作用，将雄安新区加快建设成高质量、高水平社会主义现代化城市，优化城镇规模等级结构，培育、壮大区域中心城市，构建大、中、小城市和小城镇协调发展的城镇体系。

在海洋空间安排上，优化海洋开发保护格局，构建现代化综合性港口集群，提升秦皇岛等滨海城市的服务功能和环境品质，集约高效利用海域、海岛、海岸线资源。

Territorial Spatial Planning

Territorial spatial planning is the guide for national spatial development, the spatial blueprint for sustainable development, and the basic basis for all kinds of development, protection and construction activities. On December 9, 2023, the State Council approved the *Hebei Provincial Land and Spatial Plan (2021-2035)* (hereinafter referred to as the *Plan*).

Hebei Province is located in the key area of Beijing and surrounding areas, and is one of the key regions for the implementation of the Beijing-Tianjin-Hebei coordinated development strategy. Hebei Province resolutely implements the coordinated development strategy of Beijing-Tianjin-Hebei, firmly grasps the "nose of the cattle" of relieving Beijing's non-capital functions, and jointly builds a world-class urban agglomeration with the capital as the core.

Implement the requirements of the overall pattern of coordinated development of Beijing-Tianjin-Hebei. Work with Beijing and Tianjin to promote the construction of a modern capital metropolitan area; support the linkage development of cities and counties around Beijing and Tianjin with Beijing and Tianjin; strengthen the driving role of the three development axes of Beijing-Tianjin, Beijing-Baoshi, Beijing-Tang-Qin; Promote the coordination and linkage of the four major functional areas of the core functional area around Beijing and Tianjin, the coastal first development area, the central and southern Hebei functional expansion area, and the northwest Hebei ecological conservation area; Cultivate and expand Shijiazhuang, Tangshan, Baoding, Handan and other regional central cities.

High-standard and high-quality construction of Xiong'an New Area. Facing the major national strategic needs, it effectively undertakes Beijing's non-capital functions and strives to build industrial clusters with core competitiveness; support the expansion of advantageous industries in Xiong'an New Area to the surrounding areas to form industrial clusters; Promote the interconnection of transportation infrastructure in Xiong'an New Area and its surroundings, and strengthen the basic guarantee for coordinated development.

Create a new highland for green and high-quality development in the post-Winter Olympics era. Support the development of the post-Olympic economy in Zhangbei region, strengthen the comprehensive utilization of venues after the games, and cultivate and form an international ice and snow sports and leisure tourism resort with global influence.

Build a solid spatial foundation for safe development. By 2035, the amount of cultivated land in Hebei Province will not be less than 58,673.33 square kilometers; The area of the ecological protection redline shall not be less than 36,390 square kilometers, of which the area of the marine ecological protection redline shall not be less than 1,100 square kilometers, the expansion multiple of the urban development boundary shall be controlled within 1.3 times of the scale of urban construction land based on 2020, and the area of construction land used per unit of GDP shall not be less than 41%; The retention rate of the mainland's natural coastline is not lower than that assigned by the state; The total amount of water used does not exceed the target set by the state; Except for major national projects, reclamation is completely prohibited; Strict management of uninhabited islands.

Coordinate and optimize the spatial layout of agriculture, ecology, urban and marine functions. In terms of agricultural spatial arrangement, an agricultural production pattern with concentrated agricultural production areas such as the Taihang Mountain Piedmont Plain, the Yanshan Piedmont Plain and the Heilonggang Low Plain as the main body should be constructed, so as to consolidate the spatial foundation of the granary in North China.

In terms of ecological space arrangement, we will improve the ecological functions of important areas such as the Yanshan-Taihang Mountains, the northern Beijing-Tianjin-Hebei region-Liaohe source, and the Bohai Sea coastal zone, and continue to do a good job in the construction of key ecological protection and restoration projects such as the northern sand control belt.

In terms of urban spatial arrangement, focusing on the construction of a world-class urban agglomeration in Beijing-Tianjin-Hebei, strengthening the leading role of Shijiazhuang metropolitan area, accelerating the construction of Xiong'an New Area into a high-quality and high-level socialist modern city, optimizing the scale and hierarchical structure of towns, cultivating and expanding regional central cities, and building an urban system with coordinated development of large, medium and small cities and small towns.

In terms of marine space arrangement, we will optimize the pattern of marine development and protection, build a modern comprehensive port cluster, improve the service function and environmental quality of coastal cities such as Qinhuangdao, and make intensive and efficient use of sea, island and coastline resources.

（一）严守安全底线——筑牢国家安全发展的空间基础

STRICTLY ABIDE BY THE BOTTOM LINE OF SAFETY: BUILD A SOLID FOUNDATION FOR NATIONAL SECURITY DEVELOPMENT

124.33 万平方千米 ten thousand square kilometers

319 万平方千米 ten thousand square kilometers

1.3 倍 times

坚守耕地保护"国之大者"，带位置分解下达各省（区、市）约124.33万平方千米耕地保有量和约103.07万平方千米永久基本农田保护任务，确保全国约120万平方千米耕地"实至名归"。

Adhere to the protection of arable land, which is the 'national priority', and allocate about 124.33 ten thousand square kilometers of arable land and about 103.07 ten thousand square kilometers of permanent basic farmland protection tasks to each province (region, city) based on location, to ensure that the country's about 120 ten thousand square kilometers of arable land is well deserved.

全国划定生态保护红线面积合计约319万平方千米，其中陆域生态保护红线约304万平方千米（约占陆域国土面积的31%），海洋生态保护红线不低于15万平方千米。

The national designated land ecological protection red line covers about 3.04 million square kilometers (approximately 31% of the land area); the marine ecological protection red line shall not be less than 150000 square kilometers. At the World Economic Forum held in June 2023, China was hailed.

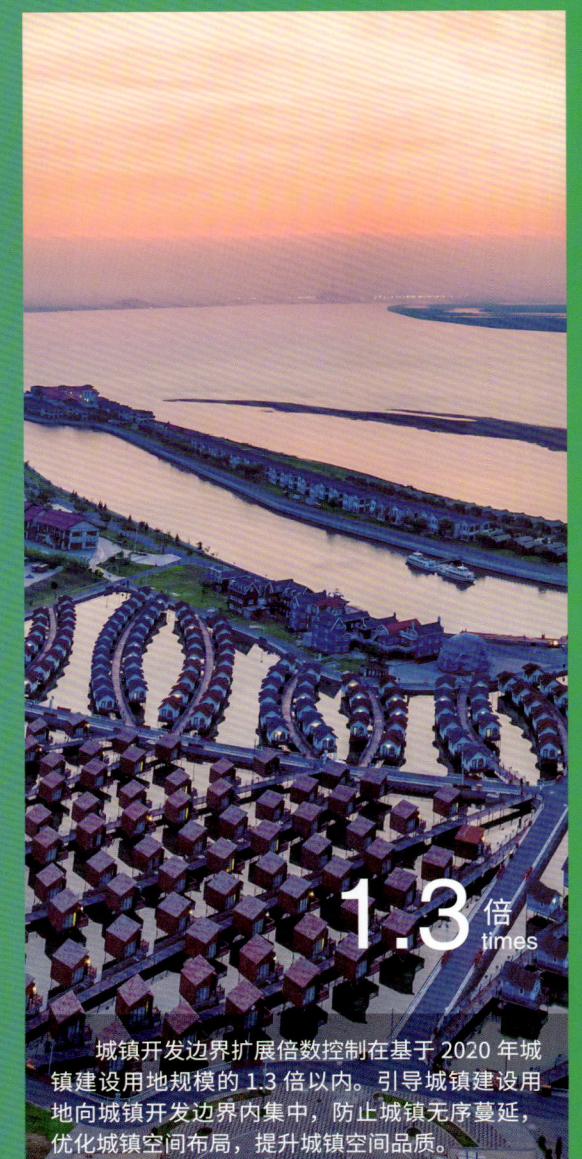

城镇开发边界扩展倍数控制在基于2020年城镇建设用地规模的1.3倍以内。引导城镇建设用地向城镇开发边界内集中，防止城镇无序蔓延，优化城镇空间布局，提升城镇空间品质。

The expansion ratio of urban development boundaries shall be controlled within 1.3 times based on the scale of urban construction land in 2020. Guide the concentration of urban construction land within the boundaries of urban development, prevent disorderly urban sprawl, optimize urban spatial layout, and enhance the quality of urban space.

（二）总体战略
OVERALL STRATEGY

底线约束 / Bottom line constraints

坚持耕地保护优先，守住粮食、生态、水、能源资源等安全底线，完善国土安全基础设施，增强空间韧性，提升国土空间可持续利用能力。

Adhere to the priority of arable land protection, guard the bottom line of food, ecology, water, energy resources and other security, improve homeland security infrastructure, enhance spatial resilience, and improve the ability to use homeland space in a sustainable manner.

集聚统筹 / Agglomeration and coordination

突出都市圈、区域中心城市和重要发展轴的引领作用，坚持"一盘棋"谋划，统筹保护和发展、陆地和海洋、近期和远期、地上和地下等空间布局，推进区域协调和城乡融合发展，优化国土空间结构。

Highlight the leading role of metropolitan areas, regional central cities and important development axes, adhere to the plan of "one game of chess", integrate the spatial layouts of protection and development, land and sea, near and far future, above and below ground, promote regional coordination and urban-rural integration, and optimize the spatial structure of the national territory.

精准协同 / Precise collaboration

牢牢牵住北京非首都功能疏解的"牛鼻子"，深入实施"两翼"带动发展战略，打造高水平承接平台，推动重点领域协同发展向纵深拓展，增强国土空间竞争力。

Firmly grasped the "nose of the cattle" of relieving Beijing's non-capital functions, deeply implement the "two-wing" driven development strategy, create high-level undertaking platforms, promote key areas of coordinated development to expand in depth, and enhance the competition of territorial space.

双向开放 / Two-way opening-up

深度融入共建"一带一路"，推动中国（河北）自由贸易试验区、雄安新区开放发展先行区等建设，推进沿海经济带高质量发展，全面提升开发区能级和水平，畅通双向开放通道，增强国土空间开放度。

Deeply integrate into the construction of the "One Belt and One Road", promote the construction of the China (Hebei) Pilot Free Trade Zone and the Xiong'an New Area, promote the high-quality development of the coastal economic belt, comprehensively upgrade the capacity and level of the development zones, unimpeded the two-way opening up of channels, and enhance the openness of the national territory and space.

集约高效 / Intensive and Efficient

改变以资源环境过度消耗为代价的增长方式，以"雄安质量"为引领构建高质量发展新动能体系，推动资源节约集约利用，提升国土空间开发效能。

Change the growth mode at the cost of excessive consumption of resources and environment, building a new kinetic energy system for high-quality development led by "Xiong'an Quality", promote the economical and intensive use of resources, and enhancing the efficiency of land space development.

品质提升 / Quality improvement

坚持以人民为中心，完善城乡生活圈，加强公共服务设施、基础设施空间保障，切实增强人民群众的获得感、幸福感、安全感，提高国土空间宜居性。

Quality improvement adheres to the people-centered approach, improves urban and rural living circles, strengthens public service facilities and infrastructure space guarantees, effectively enhances the sense of gain, security of the people, and improves the livability of the national land space.

系统治理 / System governance

强化系统观念，完善国土空间规划体系，建设国土空间规划"一张图"，充分发挥市场在资源配置中的决定性作用，更好地发挥政府作用，强化全过程公众参与，不断提高精治、共治、法治水平，推进国土空间治理体系和治理能力现代化。

Strengthen the concept of systems, improve the territorial spatial planning system, build a "one map" of territorial spatial planning, give play to the decisive role of the market in the allocation of resources, give play to the role of the government, strengthen public participation in the whole process, improve the level of rule of law, and promote the modernization of the territorial spatial governance system and governance capacity.

（三）统筹划定落实三条控制线
COORDINATE THE DELINEATION IMPLEMENTATION CONTROL LINES

到 2035 年，河北省耕地保有量不低于 58673.33 平方千米，其中永久基本农田保护面积不低于 51100 平方千米；生态保护红线面积不低于 3.639 万平方千米，其中海洋生态保护红线面积不低于 0.11 万平方千米；城镇开发边界扩展倍数控制在基于 2020 年城镇建设用地规模的 1.3 倍以内。

By 2035, the cultivated land in Hebei Province will not be less than 58,673.33 square kilometers, of which the protected area of permanent basic farmland will not be less than 51,100 square kilometers; the ecological protection red line area shall not be less than 36390 square kilometers, of which the marine ecological protection red line area shall not be less than 1100 square kilometers; the expansion ratio of urban development boundaries shall be controlled within 1.3 times based on the scale of urban construction land in 2020.

1. 优先划定耕地和永久基本农田
1. Prioritize the allocation of arable land and permanent basic farmland

现状耕地应划尽划、应保尽保，优先确定耕地保护目标，将长期利用的耕地优先划入永久基本农田，实行特殊保护。

The cultivated land under its current status should be fully demarcated and protected, and priority should be given to determining the target of cultivated land protection, and the cultivated land that has been used for a long time should be prioritized as permanent basic farmland, and special protection should be implemented.

2. 科学划定生态保护红线
2. Delineate ecological protection red line

将整合优化后的自然保护地、生态功能极脆弱区域、具有潜在重要生态价值的生态空间划入生态保护红线，统筹增强水源涵养、水土保持、防风固沙、生物多样性和海岸防护等生态系统服务功能。

Delineate the integrated and optimized nature protected areas, areas with extremely important ecological functions and extremely fragile ecological areas, and ecological and ecological spaces with potential important ecological value into the ecological protection red line, and comprehensively enhance ecosystem service functions such as water conservation, soil and water conservation, windbreak and sand fixation, biodiversity, and coastal protection.

3. 合理划定城镇开发边界
3. Delineate urban development boundaries

坚持正向约束与反向约束相结合，以城镇开发建设现状为基础，综合考虑资源承载能力、人口分布与增长趋势、城镇发展潜力，合理划定城镇开发边界。

Adhere to the combination of positive constraints and reverse constraints, based on the current situation of urban development and construction, comprehensively consider the resource carrying capacity, population distribution and growth trend, and urban development potential, and reasonably delineate the urban development boundary.

（四）完善基础设施支撑体系
IMPROVE THE INFRASTRUCTURE SUPPORT SYSTEM

落实国家重大区域基础设施建设要求，遵循高效、安全、绿色、智能的理念，在国土空间规划"一张图"上统筹协调各类基础设施布局，全面提升基础设施现代化水平和综合保障能力（见图1-1、图1-2）。

Implement the requirements of infrastructure construction in major national regions, follow the concepts of efficiency, safety, green, and intelligence, coordinate the layout of various types of infrastructure on the "one map" of territorial spatial planning, and comprehensively improve the level of infrastructure modernization and comprehensive support capabilities (see Figures 1-1 and 1-2).

交通体系 / Transportation System
- 完善"六纵六横一环"综合运输通道 / Improve the "six vertical, six horizontal and one ring" transportation channel
- 推进"轨道上的京津冀"建设 / Promote the construction of "Beijing-Tianjin-Hebei on the Rail"
- 打造现代化公路交通体系 / Build a modern highway transportation system
- 建设现代化世界级港口群 / Build a modernized world-class port cluster
- 与京津共建世界级机场群 / Build a world-class airport cluster with Beijing and Tianjin

水利设施 / Water conservancy facilities
- 河湖水系连通工程 / River and lake water system connection project
- 河道综合治理工程 / Comprehensive River Management Project
- 大中型水库工程 / Large and Medium-sized Reservoir Project
- 防洪治理工程 / Flood Control Project

能源设施 / Energy facilities
- 能源生产基地 / Energy production base
- 基础设施网络 / Infrastructure networks

新型基础设施 / New infrastructure
- 时空信息和定位导航基础设施 / Spatiotemporal information and positioning infrastructure
- 数字化基础设施 / Digital infrastructure

→ 图 1-1 基础设施支撑体系
Figure 1-1 Infrastructure support system

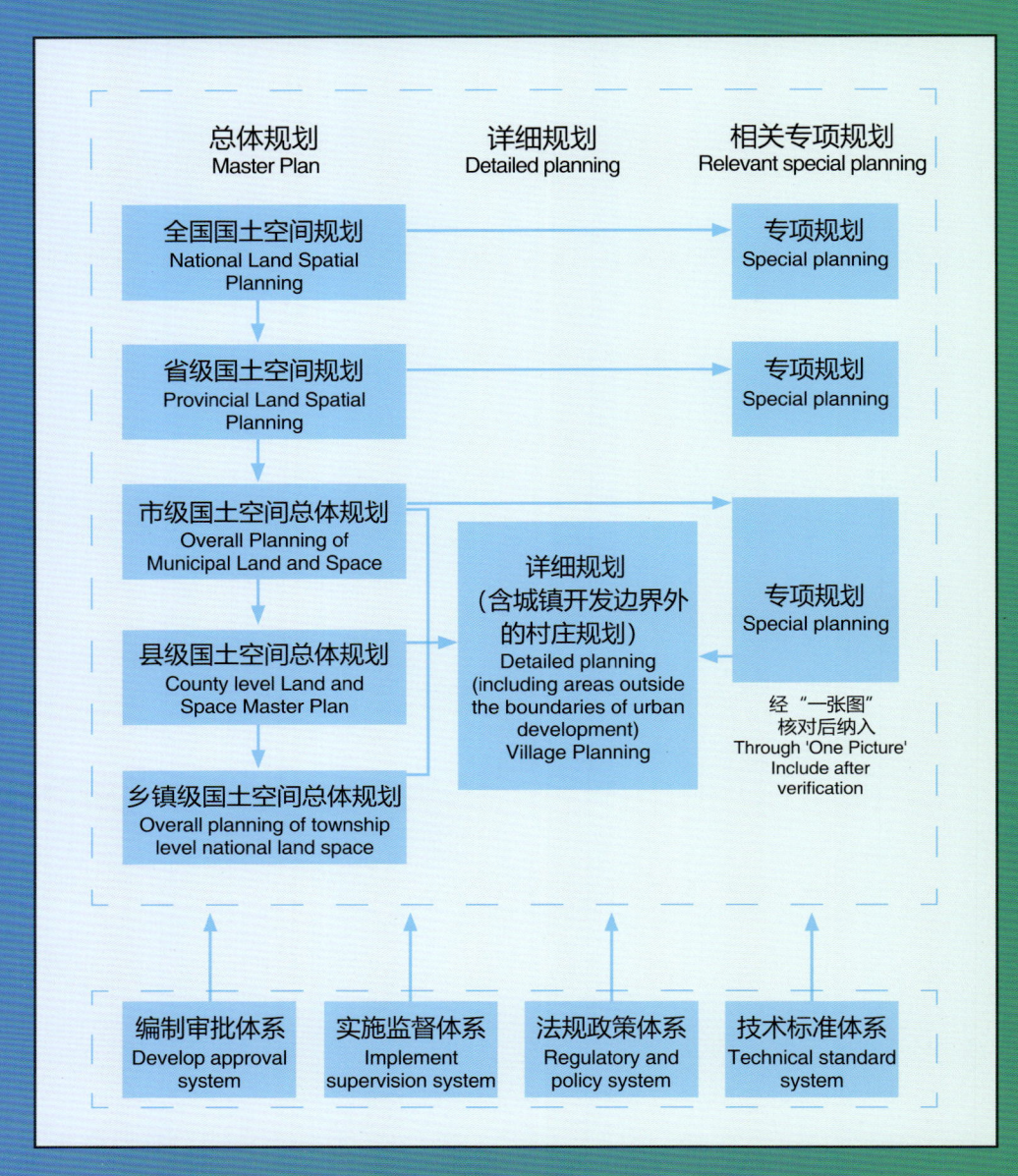

图 1-2 国土空间规划体系"五级三类四体系"
Figure 1-2 National Land Spatial Planning System "Five Levels, Three Categories, and Four Systems"

《美丽石家庄》石家庄市，摄影：杨振勇

Beautiful Shijiazhuang, Shijiazhuang City
Photographed by Yang Zhenyong

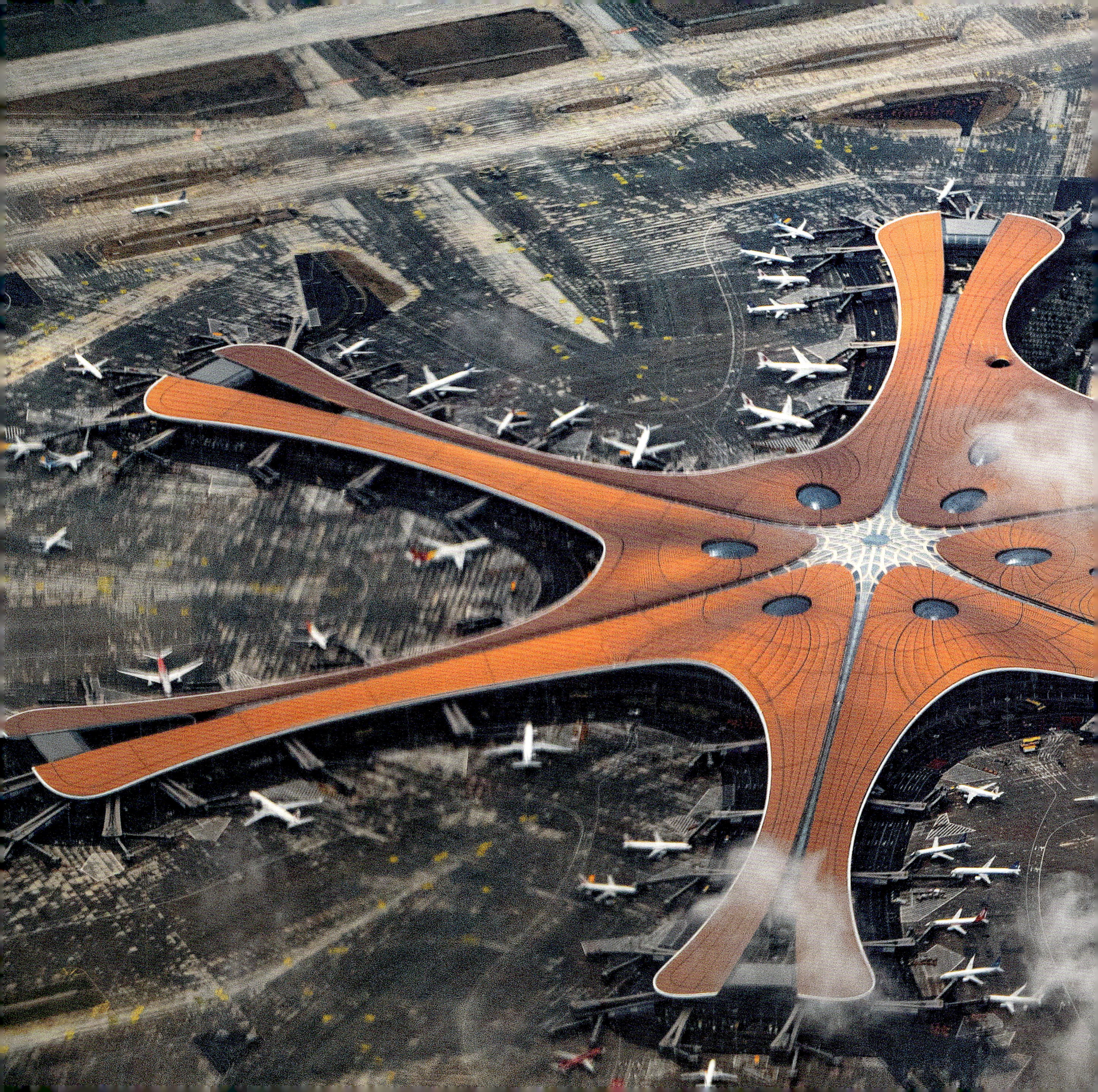

保障京津冀协同发展战略实施

落实京津冀协同发展战略要求,牢牢牵住北京非首都功能疏解的"牛鼻子",高标准、高质量建设雄安新区,深入推进京津冀协同发展,加快打造中国式现代化建设的先行区、示范区,共同构建世界一流城市群。

《壮美新国门 —— 北京大兴国际机场》,北京市大兴区,摄影:李健华

Guarantee the implementation of the coordinated development strategy of the Beijing-Tianjin-Hebei region

Implement the requirements of the coordinated development strategy of the Beijing-Tianjin-Hebei region, firmly grasped the "nose of the cattle" of relieving Beijing's non-capital functions, construct the Xiong'an New Area with high standards and quality, deeply promot the coordinated development of Beijing-Tianjin-Hebei region, accelerate the creation of a pioneer and demonstration area for the modernization of China, and jointly constructing a world-class urban agglomeration.

Magnificent New Country Gate, Beijing Daxing International Airport - Daxing District, Beijing
Photographed by Li Jianhua

高标准、高质量建设雄安新区

坚持世界眼光、国际标准、中国特色、高点定位，创造"雄安质量"，打造高质量发展的全国样板。

《雄安高铁站》，雄安新区，摄影：张学农

Construct Xiong'an New Area with High Standards and Quality

Adhere to the global vision, international standards, Chinese characteristics, and high positioning, create "Xiong'an quality" and build a national model of high-quality development.

Xiong'an High-speed Railway Station, Xiong'an New Area
Photographed by Zhang Xuenong

加强首都"两区"建设,打造后冬奥时代绿色高质量发展新高地

立足张家口与北京"山同脉、水同源、气相通"的自然生态关系,优化生态环境支撑格局,高标准规划建设张家口首都水源涵养功能区和生态环境支撑区。支持张北地区发展后奥运经济,培育形成具有全球影响力的国际冰雪运动与休闲旅游胜地。

《崇礼国家滑雪跳台中心》,张家口市崇礼区古杨树,摄影:张强

Strengthen the Construction of the "Two Zones" in the Capital City, and Create a New Green and High-quality Development in the Post Winter Olympics Era

Based on the natural ecological relationship between Zhangjiakou and Beijing, which share the same mountains, the same water and the same air, optimize the ecological environment supporting pattern, make high-standard plans to build a functional area for water conservation and an ecological environment supporting area in Zhangjiakou for the capital city; support the development of post-Olympic economy in Zhangbei county, and cultivate the formation of international ice and snow sports and leisure tourism resort with global influence.

Chongli National Ski Jumping Center, Ancient Poplar Tree in Chongli District, Zhangjiakou City
Photographed by Zhang Qiang

夯实华北粮仓空间基础

落实国家农产品主产区战略，实施"藏粮于地、藏粮于技"，坚持最严格的耕地保护制度，拓展农产品生产空间，统筹村庄空间布局，推进乡村振兴，实现由农业大省向农业强省转变。
《大丰收》，任丘市北辛庄镇，摄影：方四成

Strengthen the Spatial Foundation of North China's Granary
Implement the strategic pattern of the national main agricultural producing areas by "storing grain in the ground and storing grain in technology", adhere to the strictest arable land protection system, expand the production space of agricultural products, coordinate the spatial layout of villages, push forward the revitalization of the countryside, in order to realize the transformation from a large agricultural province to a strong agricultural province.
The Great Harvest, Beixinzhuang Town, Renqiu City
Photographed by Fang Sicheng

▎筑牢美丽河北生态屏障

建立自然保护地体系，建设生物多样性保护网络，形成以国家公园为主体、自然保护区为基础、各类自然公园为补充的自然保护地体系。
《正冠岭长城》，秦皇岛市海港区与绥中县交界处，摄影：王金磊

Building an Ecological Barrier for Beautiful Hebei
Establish a system of nature reserves, build a biodiversity protection network, and form a nature reserve system with national parks as the main body, nature reserves as the foundation, and various types of nature parks as supplements.
The Great Wall of Zhengguanling, The junction of the harbor area of Qinhuangdao City and Suizhong County
Photographed by Wang Jinlei

塑造燕赵特色魅力空间

整体保护、系统活化文化遗产和自然遗产，构建全省遗产保护空间体系，健全历史文化遗产空间保护政策机制，推动文化保护利用与农业、生态、城镇功能融合发展，打造京畿福地、老有颐养的乐享河北。

《新中国从这里走来》，石家庄市平山县西柏坡，摄影：卢白子

Shaping Charming Spaces with Yanzhao Characteristics

Overall protection and systematic revitalization of cultural and natural heritage, build a spatial system for heritage protection in the entire province, sound the policy mechanism for spatial protection of historical and cultural heritage, promote the integrated development of cultural protection and utilization with agricultural, ecological and urban functions, and create a blessed place in Beijing and its surrounding areas and a pleasant Hebei where the elderly can take care of themselves.

New China Comes from Here, Xibaipo, Pingshan County, Shijiazhuang City

Photographed by Lu Baizi

构筑向海图强战略要地

坚持向海发展、向海图强，经略蓝色国土，科学开发海洋资源，保护海洋生态环境，优化沿海地区空间布局，推进港口转型升级和资源整合，建成临港产业强省，打造河北高质量发展战略要地。

《唐山曹妃甸港》，唐山市曹妃甸港，摄影：李克东

Construct a Strategic Location for Seaward Development

Adhere to the development and strength of the sea, strategize the blue land, scientifically develop marine resources, protect the marine ecological environment, optimize the spatial layout of the coastal area, promote the transformation and upgrading of ports and resource integration, build a strong province of port industry, and create a strategic location for the high-quality development of Hebei.

Port of Caofeidian, Tangshan, Port of Caofeidian, Tangshan City
Photographed by Li Kedong

土地资源

河北被称为"浓缩的国家地理读本",是我国唯一一个集齐所有地貌的省份。河北地处华北大地,地势为西北高、东南低,从西北向东南依次为坝上高原、燕山和太行山地、河北平原三大地貌单元,并呈半环状逐级下降。

土地是经济社会发展最重要的物质基础。河北不断完善节约集约用地措施,提升土地资源利用效率,在有限的土地上实现更高水平利用。截至2023年,河北25个县(市)被认定为国家级自然资源集约示范县(市),为全国最多。

截至2022年年底,河北省共有耕地面积约60112平方千米、园地约9699.60平方千米、林地约64180.47平方千米、草地约19200.73平方千米、湿地约7268平方千米、城镇村及工矿用地约21547.53平方千米、交通运输用地约4256.53平方千米、水域及水利设施用地约5868.33平方千米。详情见图2-1至图2-10。

粮安天下,地为根基。河北耕地资源充足,主要分布在平原区和坝上地区。沧州、保定、张家口三市耕地面积较大,约占全省的40%。耕地类型主要有水田、水浇地和旱地,其中水田约有962.07平方千米,占耕地面积的1.60%;水浇地约有38564.33平方千米,占耕地面积的64.15%;旱地约有20585.67平方千米,占耕地面积的34.25%。

截至2022年底,河北已建成高标准农田约34880平方千米,占全省耕地面积的57.8%。

保定市是京津冀最重要的"菜篮子""米袋子"基地。邯郸市形成了"叠石相次,包土成田"旱作石堰梯田系统,总面积约178.67平方千米,被联合国世界粮食计划署专家称为"中国第二长城",是全球重要的农业文化遗产之一。塞北梯田位于张家口崇礼区腹地,被称为"距离北京最近的梯田"。

河北还有约3866.67平方千米盐碱地,其中盐碱耕地3800多平方千米,是我国盐碱地面积较大的省份之一。

Land Resources

Hebei is known as "The Condensed National Geographic Reader" and is the only province in China that has collected all the landforms. Hebei is located in the land of North China, the terrain is high in the northwest and low in the southeast. From the northwest to the southeast, there are three major geomorphological units: Bashang Plateau, Yanshan and Taihang Mountains, and Hebei Plain, which are descending in a semi-circular shape.

By the end of 2022, Hebei Province had a total of about 60,112 square kilometers of cultivated land, about 9,699.60 square kilometers of orchard land, about 64,180.47 square kilometers of forest land, about 19,200.73 square kilometers of grassland, about 7,268 square kilometers of wetland, about 21,547.53 square kilometers of urban and village and industrial and mining land, about 4,256.53 square kilometers of transportation land, and about 5,868.33 square kilometers of water and water conservancy facilities. See Figure 2-1 to Figure 2-10 for details.

Food security in the world, the earth is the foundation. Hebei has sufficient cultivated land resources, mainly distributed in the plain area and Bashang area, and the cultivated land area of Cangzhou, Baoding and Zhangjiakou is relatively large, accounting for about 40% of the whole province. The main types of cultivated land are paddy field, irrigated land and dry land, of which about 962.07 square kilometers are paddy fields, accounting for 1.60% of the cultivated land. about 38,564.33 square kilometers of irrigated land, accounting for 64.15% of cultivated land; about 20,585.67 square kilometers of dry land, accounting for 34.25% of cultivated land.

By the end of 2022, about 34,880 square kilometers of high-standard farmland had been built in Hebei, accounting for 57.8% of the province's cultivated land.

Baoding City is the most important "vegetable basket" and "rice bag" base in Beijing-Tianjin-Hebei. Handan City has formed a dry stone weir terrace system of "stacked stones, covered with soil into fields", with a total area of about 178.67 square kilometers, which is called "China's second Great Wall" by experts of the United Nations World Food Programme, and is one of the world's important agricultural cultural heritages. Located in the hinterland of Chongli District, Zhangjiakou, the Saibei Rice Terraces are known as the "Terraced Rice Fields Closest to Beijing".

Hebei also has about 3,866.67 square kilometers of saline-alkali land, of which more than 3,800 square kilometers of saline-alkali cultivated land, and is one of the provinces with a large area of saline-alkali land in China.

Land is the most important material foundation for economic and social development. Hebei has continuously improved measures to save and intensively use land, improve the efficiency of land resource utilization, and achieve a higher level of utilization on limited land. As of 2023, 25 counties (cities) in Hebei have been recognized as national-level demonstration counties (cities) for intensive natural resource conservation, the most in the country.

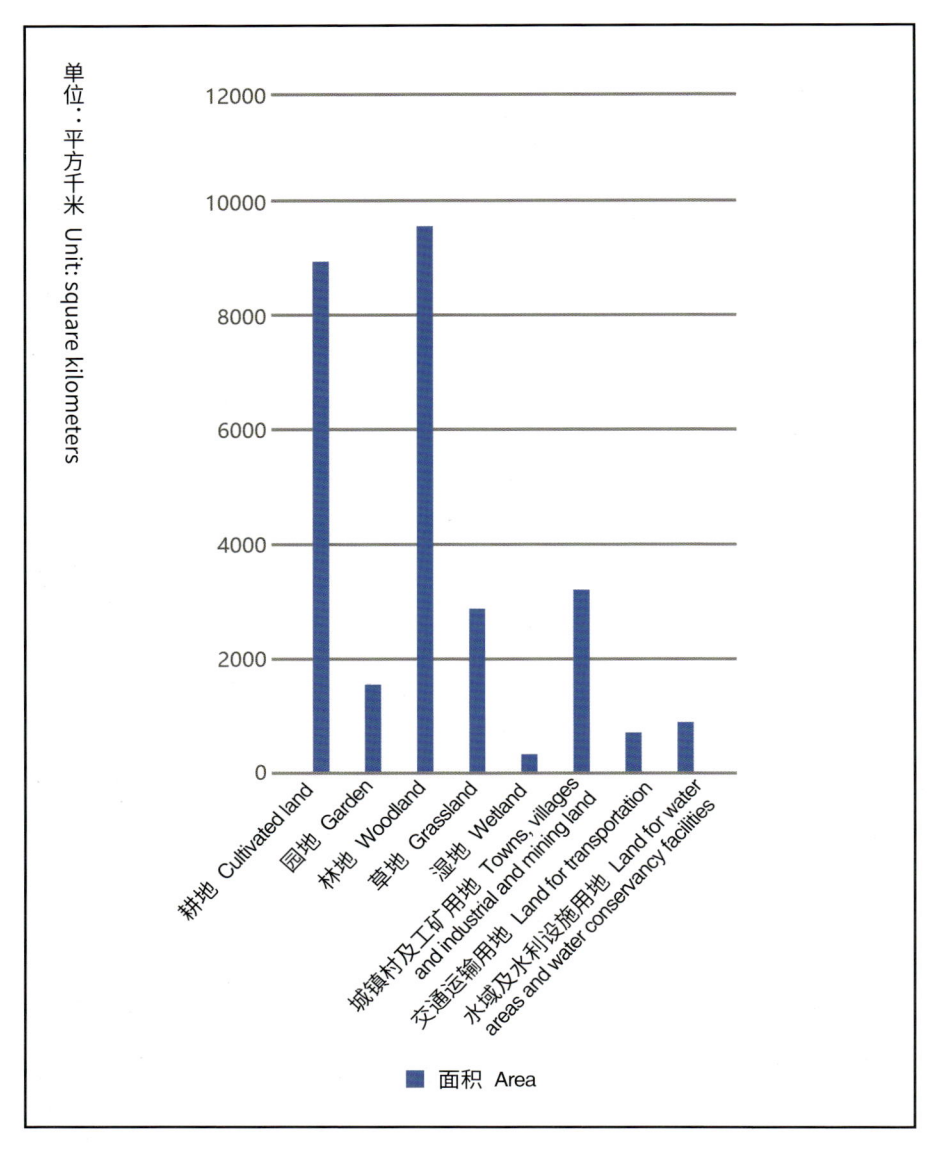

图 2-1 2022 年河北省土地资源概况
Figure 2-1 Overview of land resources in Hebei Province in 2022

《太行高速——破峪成龙》，太行山脉，摄影：樊高瑞

Taihang Expressway - Breaking the Valley Like a Dragon
Taihang Mountains
Photographed by Fan Gaorui

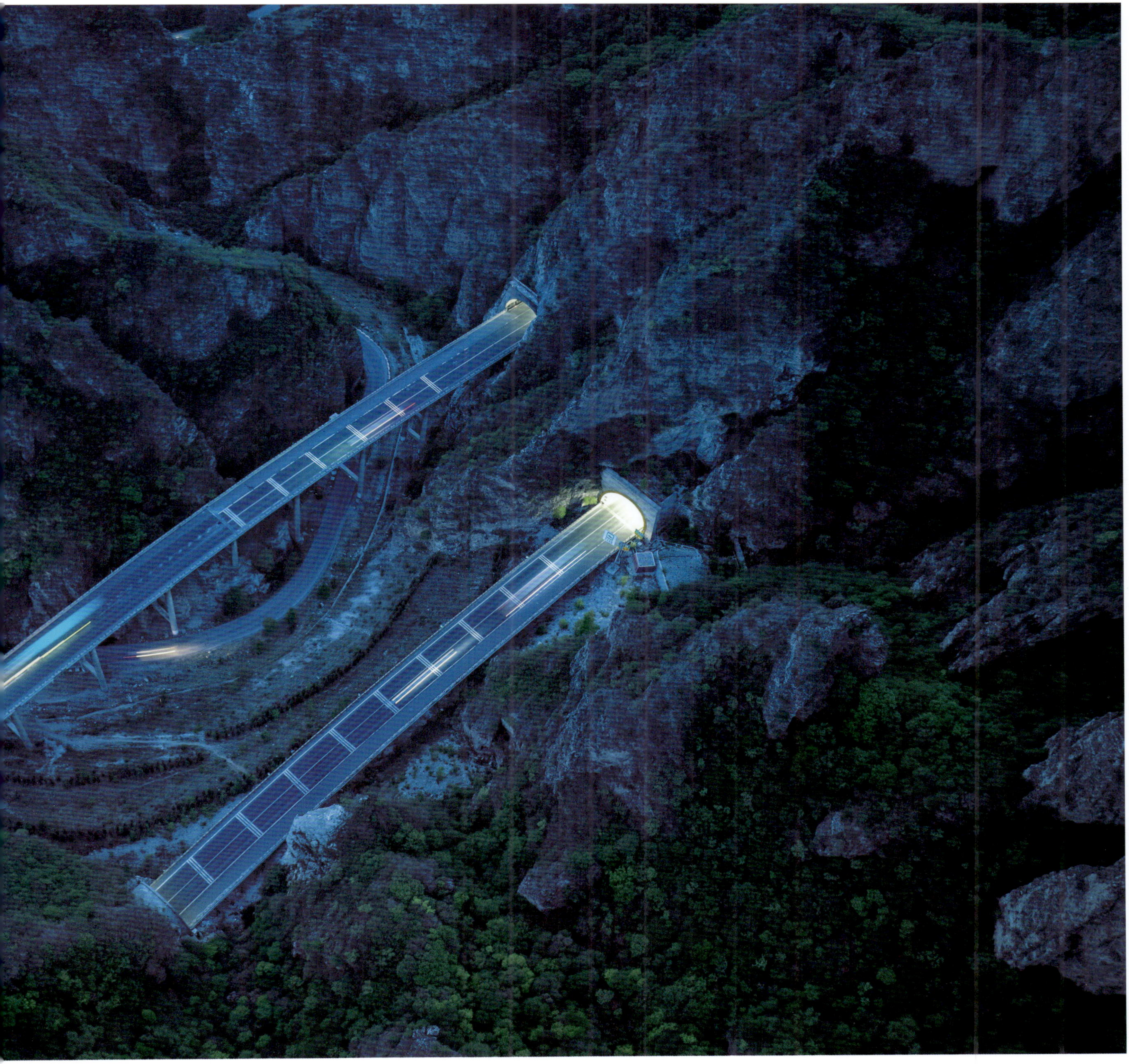

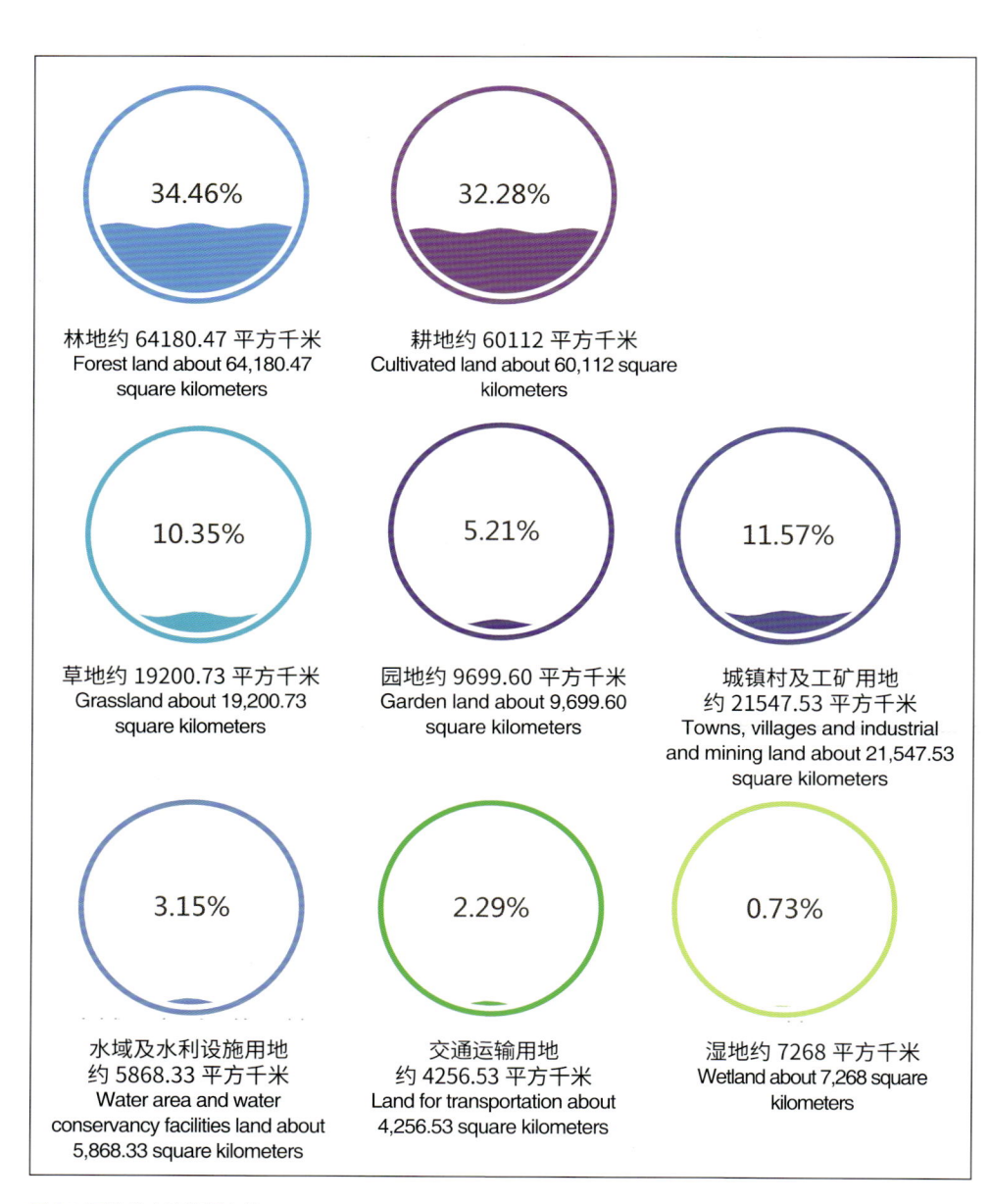

34.46% 林地约 64180.47 平方千米 Forest land about 64,180.47 square kilometers	**32.28%** 耕地约 60112 平方千米 Cultivated land about 60,112 square kilometers	
10.35% 草地约 19200.73 平方千米 Grassland about 19,200.73 square kilometers	**5.21%** 园地约 9699.60 平方千米 Garden land about 9,699.60 square kilometers	**11.57%** 城镇村及工矿用地约 21547.53 平方千米 Towns, villages and industrial and mining land about 21,547.53 square kilometers
3.15% 水域及水利设施用地约 5868.33 平方千米 Water area and water conservancy facilities land about 5,868.33 square kilometers	**2.29%** 交通运输用地约 4256.53 平方千米 Land for transportation about 4,256.53 square kilometers	**0.73%** 湿地约 7268 平方千米 Wetland about 7,268 square kilometers

图 2-2 河北省土地资源占比

Figure 2-2 Proportion of land resources in Hebei Province

《大地音符涌春潮》，保定市，摄影：张建宇

Spring Bursts Through the Fields, Baoding City
Photographed by Zhang Jianyu

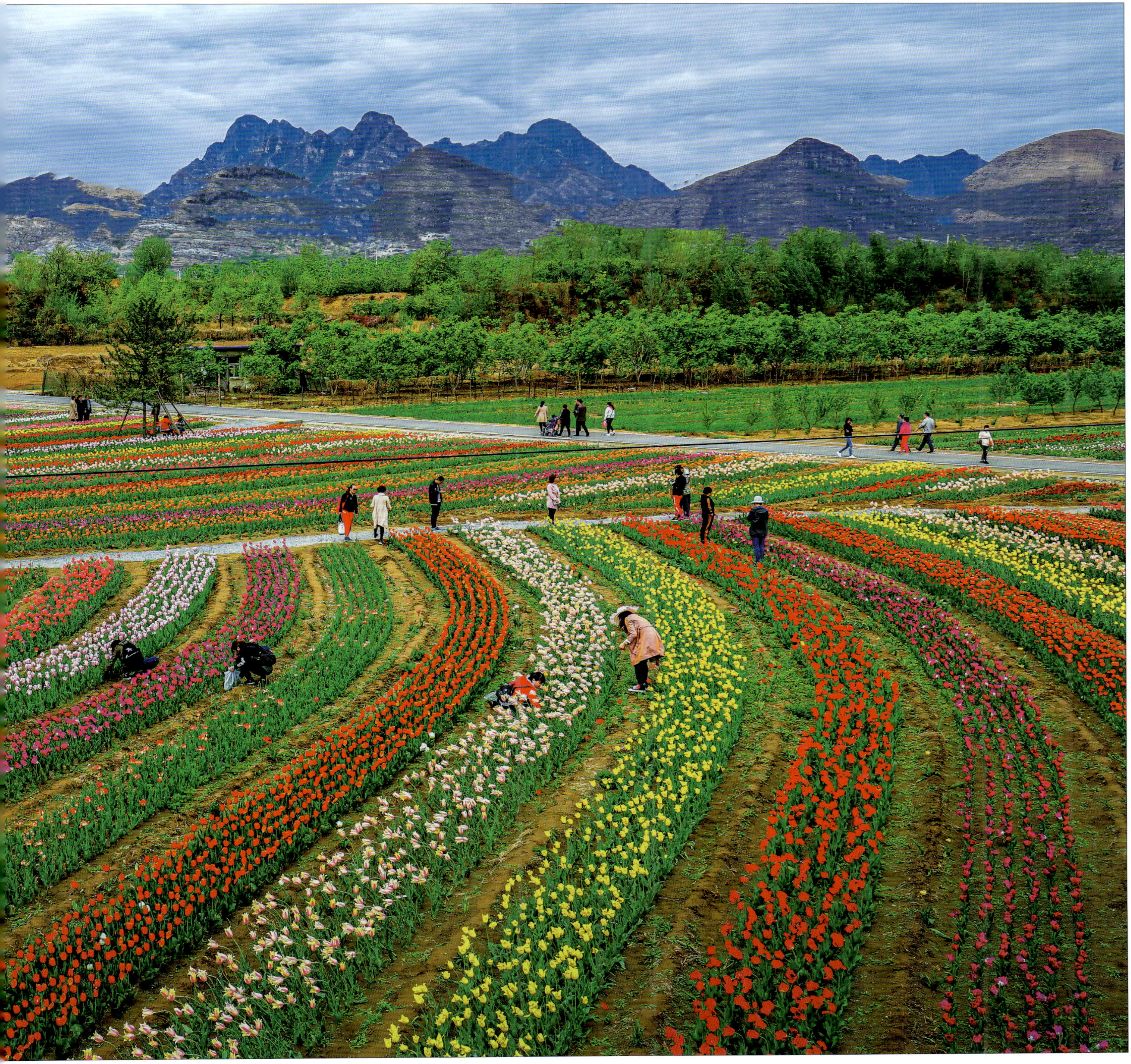

■ 粮食

　　河北的耕地主要种植小麦、玉米、大豆、棉花等农作物，一年两熟地区的耕地约占全省耕地的 63.71%。
《风吹麦浪》，保定市定兴县，摄影：张建宇

Food
Cultivated land in Hebei mainly grows wheat, maize, soybeans, cotton and other crops, of which 63.71% is cultivated in biannual areas.
The Wind Blows the Wheat Waves, Dingxing County, Baoding City
Photographed by Zhang Jianyu

园地

河北省园地面积约为 9699.6 平方千米。其中果园约有 9395.87 平方千米，占 96.87%；其他园地约有 303.53 平方千米，占 0.51%；另有茶园约 0.2 平方千米。
《春风十里杏花路》，保定市顺平县，摄影：张建宇

Garden land
The area of garden land in Hebei Province is approximately 9,699.6 square kilometers. Among them, the orchard covers an area of approximately 9,395.87 square kilometers, accounting for 96.87% of the total; other gardens cover an area of approximately 303.53 square kilometers, accounting for 0.51%; and there is also a tea garden of about 0.2 square kilometers.
Ten Miles of Blooming Apricot Road in the Spring Breeze, Shunping County, Baoding City
Photographed by Zhang Jianyu

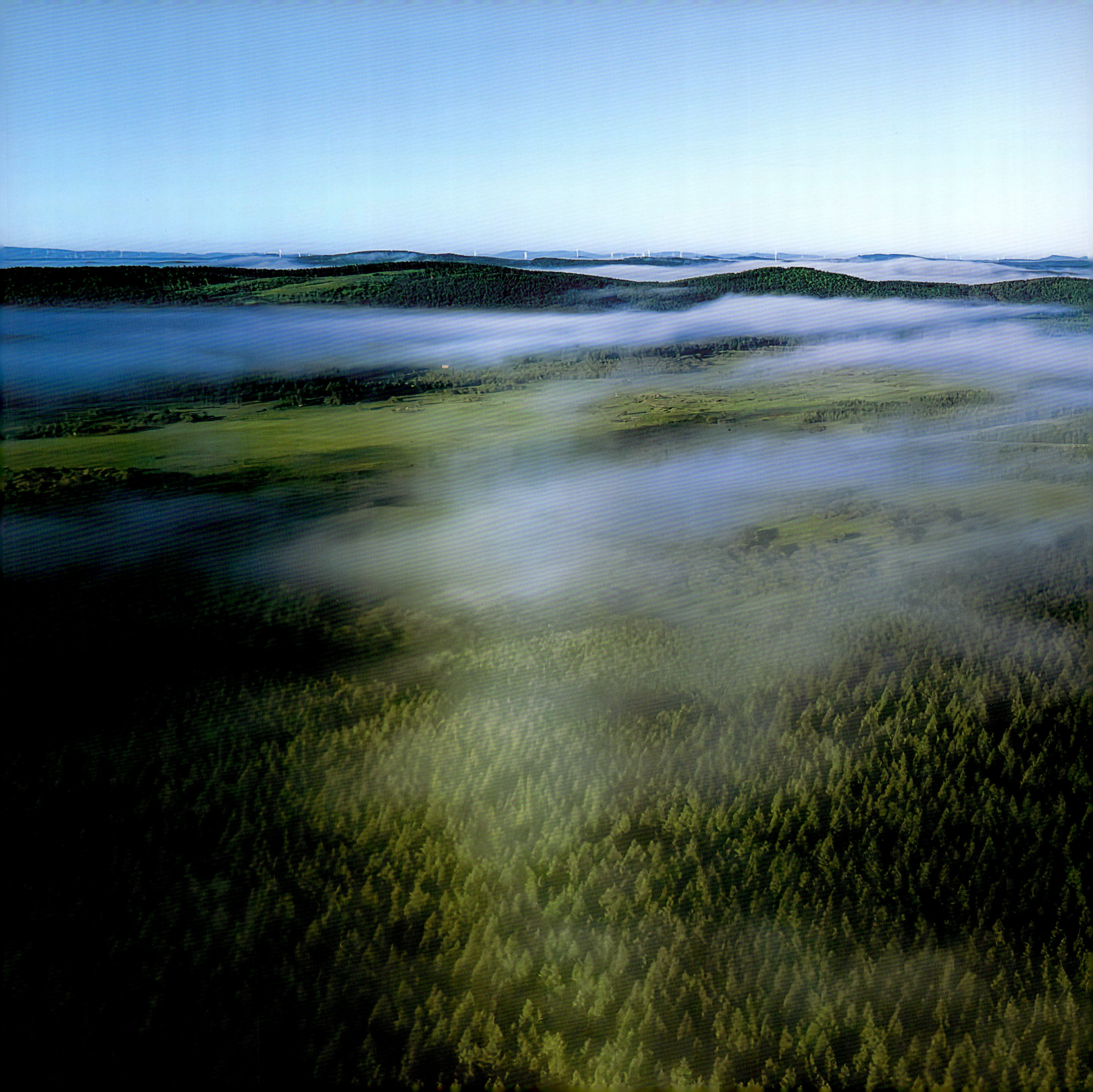

■ 林地

　　河北省约有林地面积 64180.47 平方千米。其中乔木林地面积约为 26414.67 平方千米，占 41.16%；灌木林地面积约为 19856.13 平方千米，占 30.94%；其他林地面积约为 17909.40 平方千米，占 27.90%；另有竹林地约 0.2 平方千米。
《塞罕坝机械林场》，承德市围场县，摄影：韩宏亮

Forest land
Hebei Province has a forest area of approximately 64,180.47 square kilometers. The area of deciduous forest land is approximately 26,414.67 square kilometers, accounting for 41.16%; shrub covers about 19,856.13 square kilometers, accounting for 30.94%; other covers about 17,909.40 square kilometers, accounting for 27.90%; and bamboo covers about 0.2 square kilometers.
Saihanba Mechanical Forest Farm, Weichang County, Chengde City
Photographed by Han Hongliang

■ 草地

　　河北省约有草地面积 19200.73 平方千米。其中天然牧草地面积约为 4119.13 平方千米，占 21.45%；人工牧草地约有 106.13 平方千米，占 0.55%；其他草地约有 14975.47 平方千米，占 77.99%。
《坝上草原》，承德市丰宁县坝上，摄影：刘云华

Grassland
Hebei Province has a grassland area of approximately 19,200.73 square kilometers, of which the natural pastureland area is approximately 4,119.13 square kilometers, accounting for 21.45%; the artificial pastureland covers an area of approximately 106.13 square kilometers, accounting for 0.55%; and there are approximately 14,975.47 square kilometers of other grasslands, accounting for 77.99%.
Bashang Grassland, On the dam of Fengning County, Chengde City
Photographed by Liu Yunhua

湿地

　　河北省约有湿地面积 7268 平方千米。其中沿海滩涂面积约为 627.40 平方千米，占 45.89%；内陆滩涂约有 477.67 平方千米，占 34.94%；沼泽草地约有 171.60 平方千米，占 12.55%；沼泽地约有 79.40 平方千米，占 5.81%；灌丛沼泽约有 10.53 平方千米，占 0.77%；森林沼泽约有 0.467 平方千米，占 0.04%。

《白洋淀湿地》，保定市白洋淀，摄影：薛志勇

Wetland

Hebei Province has an area of about 7,268 square kilometers. Among them, the area along the beach is about 627.40 square kilometers, accounting for 45.89%; The inland tidal flat is about 477.67 square kilometers, accounting for 34.94%; The swamp grassland is about 171.60 square kilometers, accounting for 12.55%; The marshland is about 79.40 square kilometers, accounting for 5.81%; The shrub swamp is about 10.53 square kilometers, accounting for 0.77%; The forest swamp is about 0.467 square kilometers, or 0.04%.

Baiyangdian Wetland, Baiyangdian, Baoding City
Photographed by Xue Zhiyong

城镇村及工矿用地

　　河北省城镇村及工矿用地面积约有 21547.53 平方千米，其中城市用地面积约为 2037.40 平方千米，占 9.46%；建制镇用地面积约为 3645.13 平方千米，占 16.92%；村庄用地面积约为 13503 平方千米，占 62.67%；采矿用地面积约为 1979 平方千米，占 9.18%；风景名胜及特殊用地面积约为 383 平方千米，占 1.78%。
《美丽乡村》，张家口市崇礼区，摄影：方四成

Urban Villages and Industrial and Mining Land
The area of urban villages and industrial and mining land in Hebei Province is about 21,547.53 square kilometers, of which the urban land area is about 2,037.40 square kilometers, accounting for 9.46%; The land area of the town is about 3,645.13 square kilometers, accounting for 16.92%; The land area of the village is about 13,503 square kilometers, accounting for 62.67%; The area of mining land is about 1,979 square kilometers, accounting for 9.18%; The area of scenic spots and special land is about 383 square kilometers, accounting for 1.78%.
Beautiful Countryside, Chongli District, Zhangjiakou City
Photographed by Fang Sicheng

水域及水利设施用地

河北省水域及水利设施用地面积约有 5868.33 平方千米，其中河流水面面积约为 1797.13 平方千米，占 30.62%；湖泊水面面积约为 252.27 平方千米，占 4.30%；水库水面面积约为 720.80 平方千米，占 12.28%；坑塘水面面积约为 1409.67 平方千米，占 24.02%；沟渠面积约为 1319.80 平方千米，占 22.49%；水工建筑用地约 368.67 平方千米，占 6.28%。

《朱庄水库》，沙河市綦村镇朱庄村，摄影：冯九伟

Water Area and Water Conservancy Facilities Land

Hebei Province has a land area of water area and water conservancy facilities about 5,868.33 square kilometers, of which the water surface area of rivers is about 1,797.13 square kilometers, accounting for 30.62%; The water surface area of the lake is about 252.27 square kilometers, accounting for 4.30%; The water surface area of the reservoir is about 720.80 square kilometers, accounting for 12.28%; The water surface area of the pit pond is about 1409.67 square kilometers, accounting for 24.02%; The area of the ditch is about 1319.80 square kilometers, accounting for 22.49%; The land for hydraulic construction is about 368.67 square kilometers, accounting for 6.28%.

Zhuzhuang Reservoir, Zhuzhuang Village, Qicun Town, Shahe City
Photographed by Feng Jiuwei

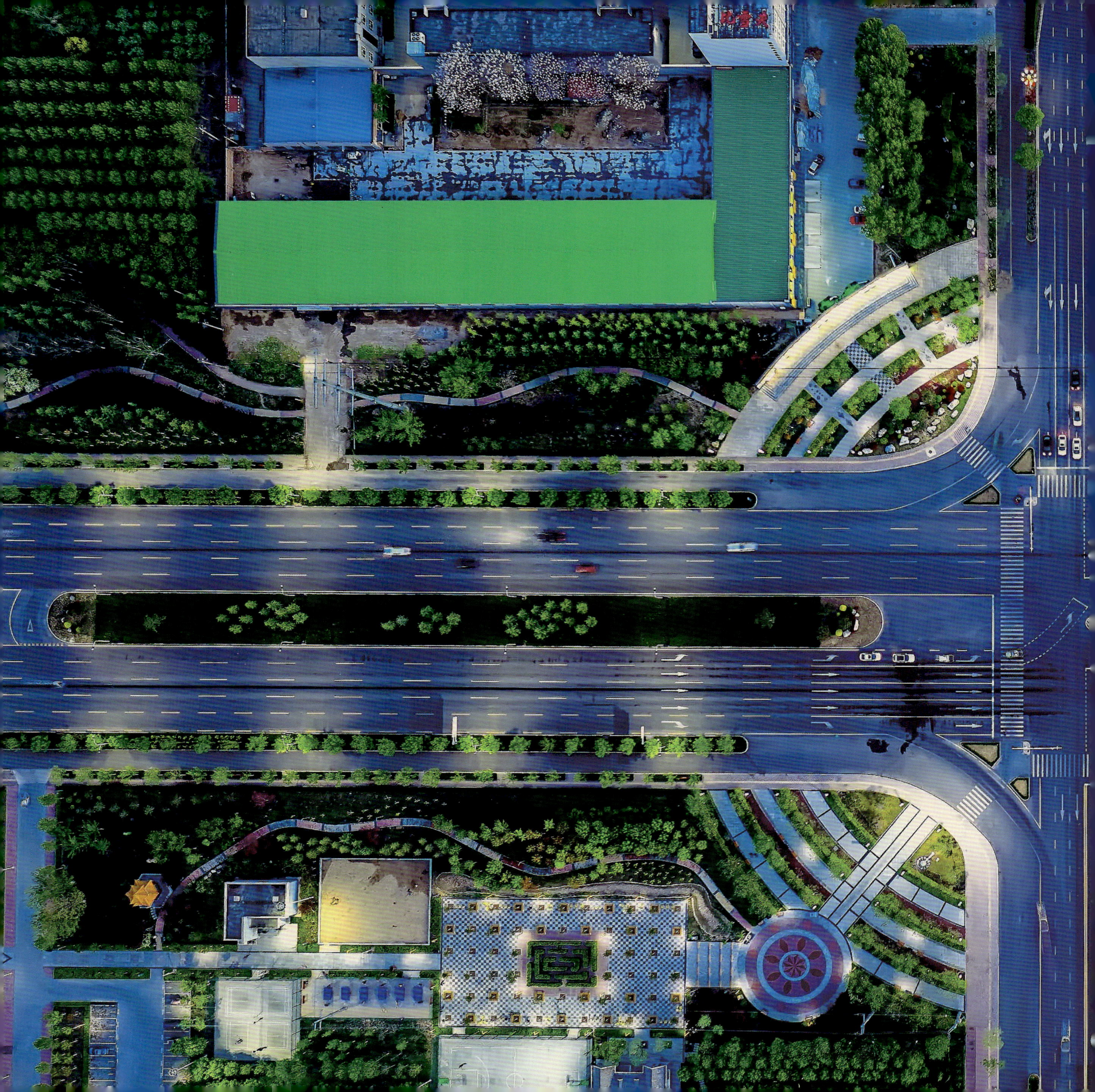

交通运输用地

河北省交通运输用地面积约有 4256.53 平方千米，其中铁路用地面积约为 333.67 平方千米，占 7.84%；轨道交通用地面积约为 2.67 平方千米，占 0.06%；公路用地面积约为 1940.07 平方千米，占 45.58%；农村道路面积约为 1893.20 平方千米，占 44.48%；机场用地面积约为 30.60 平方千米，占 0.72%；港口码头用地面积约为 54.53 平方千米，占 1.28%；管道运输用地面积约为 1.80 平方千米，占 0.04%。

《蓝图》，辛集市澳森大街，摄影：种勇

Transportation Land

Hebei Province has about 4,256.53 square kilometers of transportation land, of which 333.67 square kilometers are railway land, accounting for 7.84%; The land area of rail transit is about 2.67 square kilometers, accounting for 0.06%; The land area of highways is about 1,940.07 square kilometers, accounting for 45.58%; The area of rural roads is about 1,893.20 square kilometers, accounting for 44.48%; The land area of the airport is about 30.60 square kilometers, accounting for 0.72%; The land area of the port terminal is about 54.53 square kilometers, accounting for 1.28%; The land area for pipeline transportation is about 1.80 square kilometers, accounting for 0.04%.

Blueprint, Aosen Street, Xinji City
Photographed by Chong Yong

各类细分地类图表
CHART OF THE VARIOUS SUBDIVIDED LAND CATEGORIES

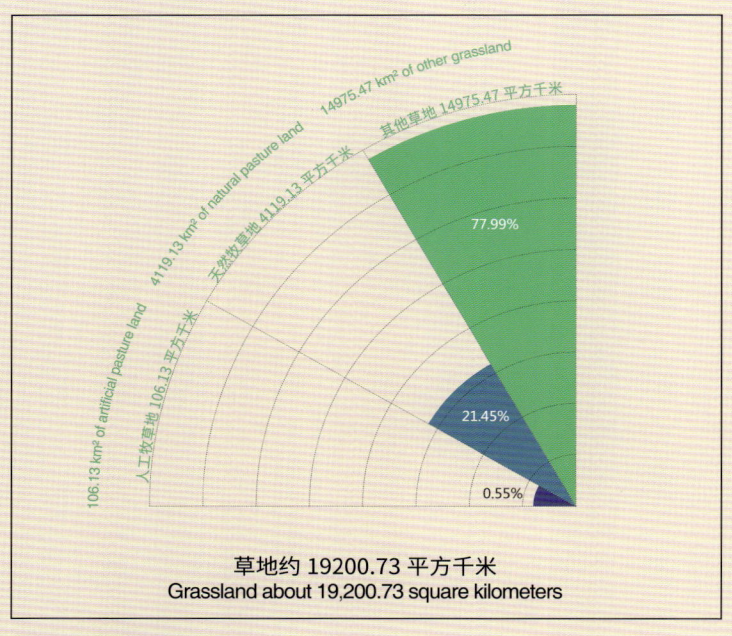

草地约 19200.73 平方千米
Grassland about 19,200.73 square kilometers

- 其他草地 14975.47 平方千米 (77.99%) — 14975.47 km² of other grassland
- 天然牧草地 4119.13 平方千米 (21.45%) — 4119.13 km² of natural pasture land
- 人工牧草地 106.13 平方千米 (0.55%) — 106.13 km² of artificial pasture land

图 2-3 草地细分比重图
Figure 2-3 Proportion of different types of grassland in subdivision

湿地约 7268 平方千米
Wetland about 7,268 square kilometers

- 沿海滩涂 627.40 平方千米 45.89% — 627.40 km² along the beach
- 内陆滩涂 477.67 平方千米 34.94% — 477.67 km² of inland tidal flats
- 沼泽草地 171.60 平方千米 12.55% — 171.60 km² of swamp grassland
- 沼泽地 79.40 平方千米 5.81% — marshland 79.40 km²
- 灌丛沼泽 10.53 平方千米 0.77% — shrub swamp 10.53 km²
- 森林沼泽 0.467 平方千米 0.04% — The forest swamp is about 0.467 km²

图 2-4 湿地细分比重图
Figure 2-4 Proportion of different types of wetlands

水域及水利设施用地约有 5868.33 平方千米
Water area and water conservancy facilities land about 5,868.33 square kilometers

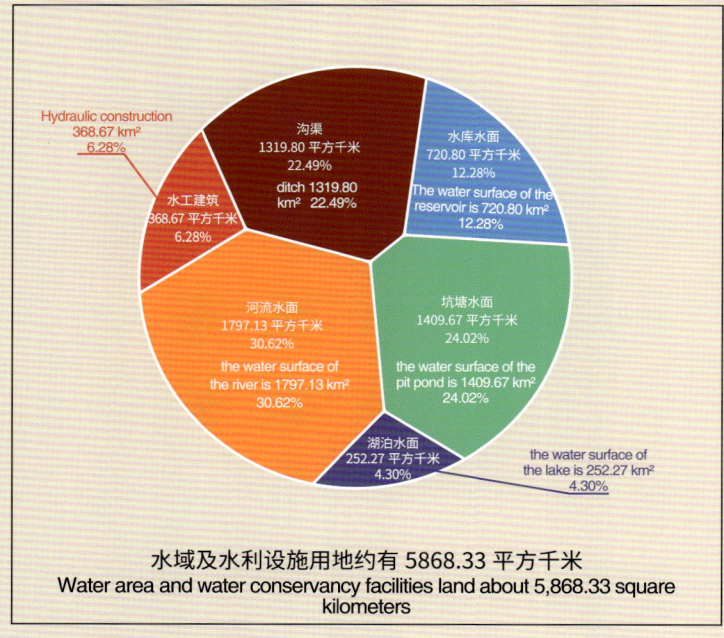

- 沟渠 1319.80 平方千米 22.49% — ditch 1319.80 km² 22.49%
- 水库水面 720.80 平方千米 12.28% — The water surface of the reservoir is 720.80 km²
- 河流水面 1797.13 平方千米 30.62% — the water surface of the river is 1797.13 km² 30.62%
- 坑塘水面 1409.67 平方千米 24.02% — the water surface of the pit pond is 1409.67 km²
- 湖泊水面 252.27 平方千米 4.30% — the water surface of the lake is 252.27 km² 4.30%
- 水工建筑 368.67 平方千米 6.28% — Hydraulic construction 368.67 km²

图 2-7 水域及水利设施用地细分比重图
Figure 2-7 Proportion of land used for different types of water areas and water conservancy facilities

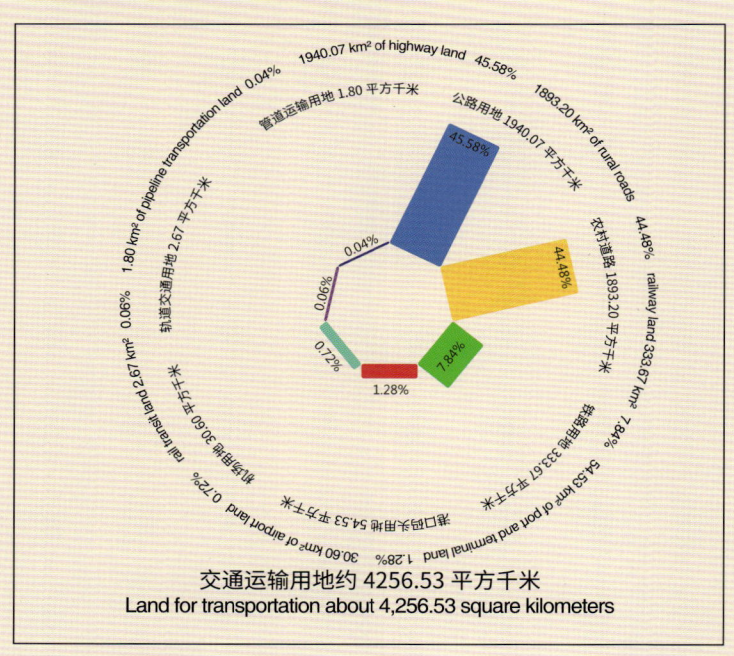

交通运输用地约 4256.53 平方千米
Land for transportation about 4,256.53 square kilometers

- 公路用地 1940.07 平方千米 45.58% — 1940.07 km² of highway land
- 农村道路 1893.20 平方千米 44.48% — 1893.20 km² of rural roads
- 铁路用地 333.67 平方千米 7.84% — railway land 333.67 km²
- 港口码头用地 54.53 平方千米 1.28% — 54.53 km² of port and terminal land
- 机场用地 30.60 平方千米 0.72% — 30.60 km² of airport land
- 轨道交通用地 2.67 平方千米 0.06% — rail transit land 2.67 km²
- 管道运输用地 1.80 平方千米 0.04% — 1.80 km² of pipeline transportation land

图 2-8 交通运输用地细分比重图
Figure 2-8 Breakdown of different types of transportation land

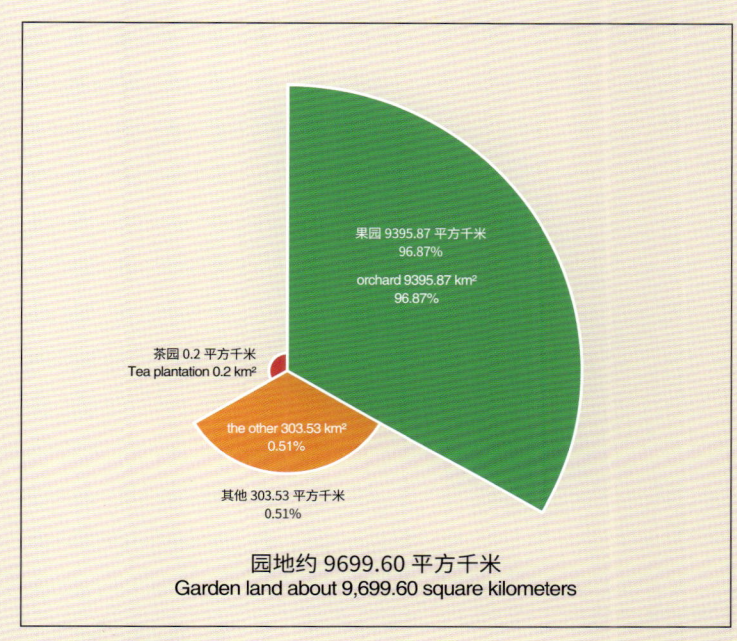

图 2-5 园地细分比重图
Figure 2-5 Proportion of different types of garden land

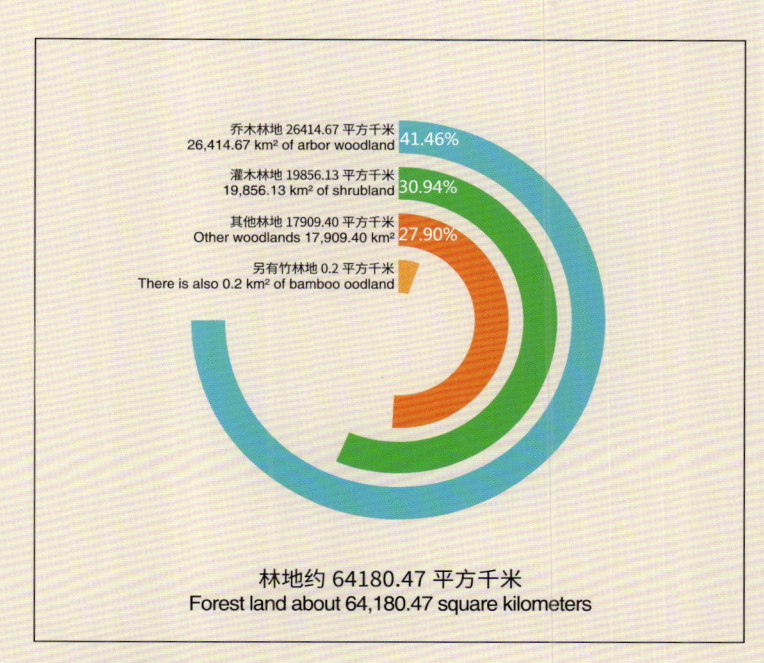

图 2-6 林地细分比重图
Fig.2-6 Proportion of different types of forest land

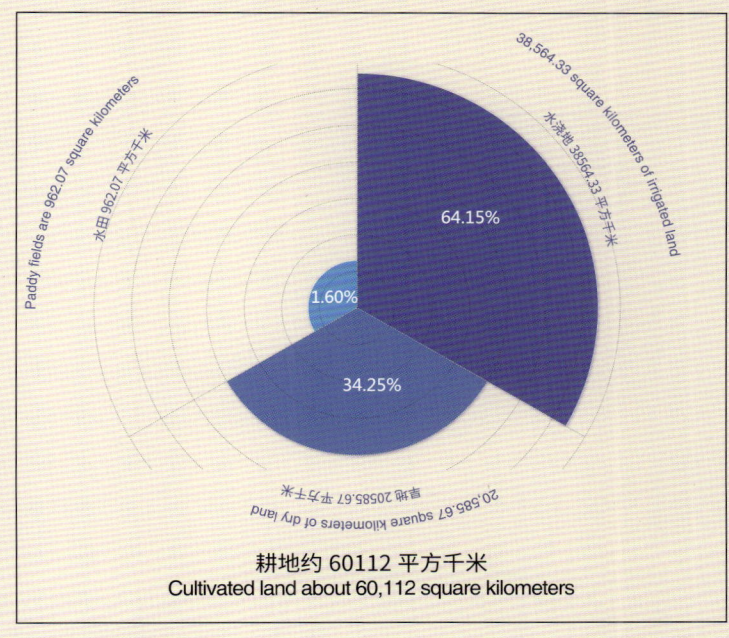

图 2-9 耕地细分比重图
Figure 2-9 Proportion of different types of cultivated land

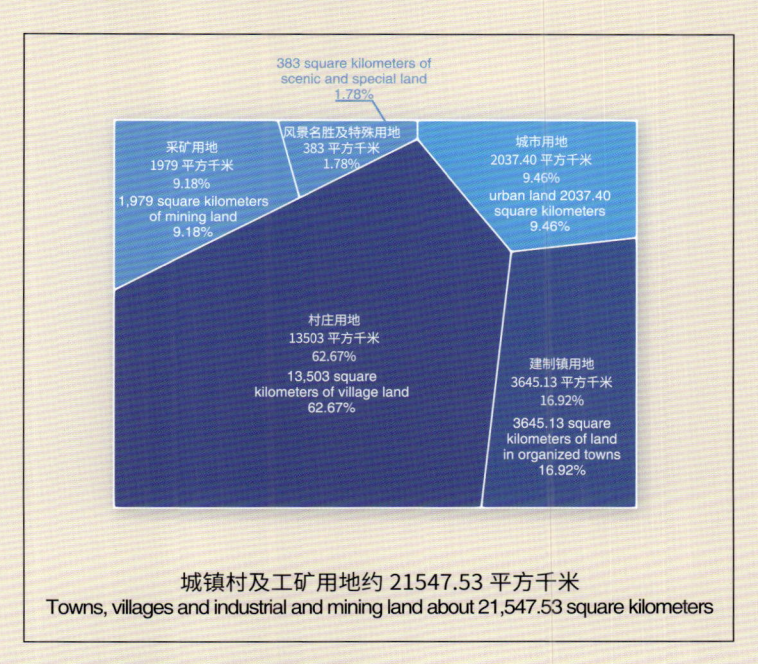

图 2-10 城镇村及矿工用地细分比重图
Figure 2-10 Proportion of different types of Towns, villages and industrial and mining land

063

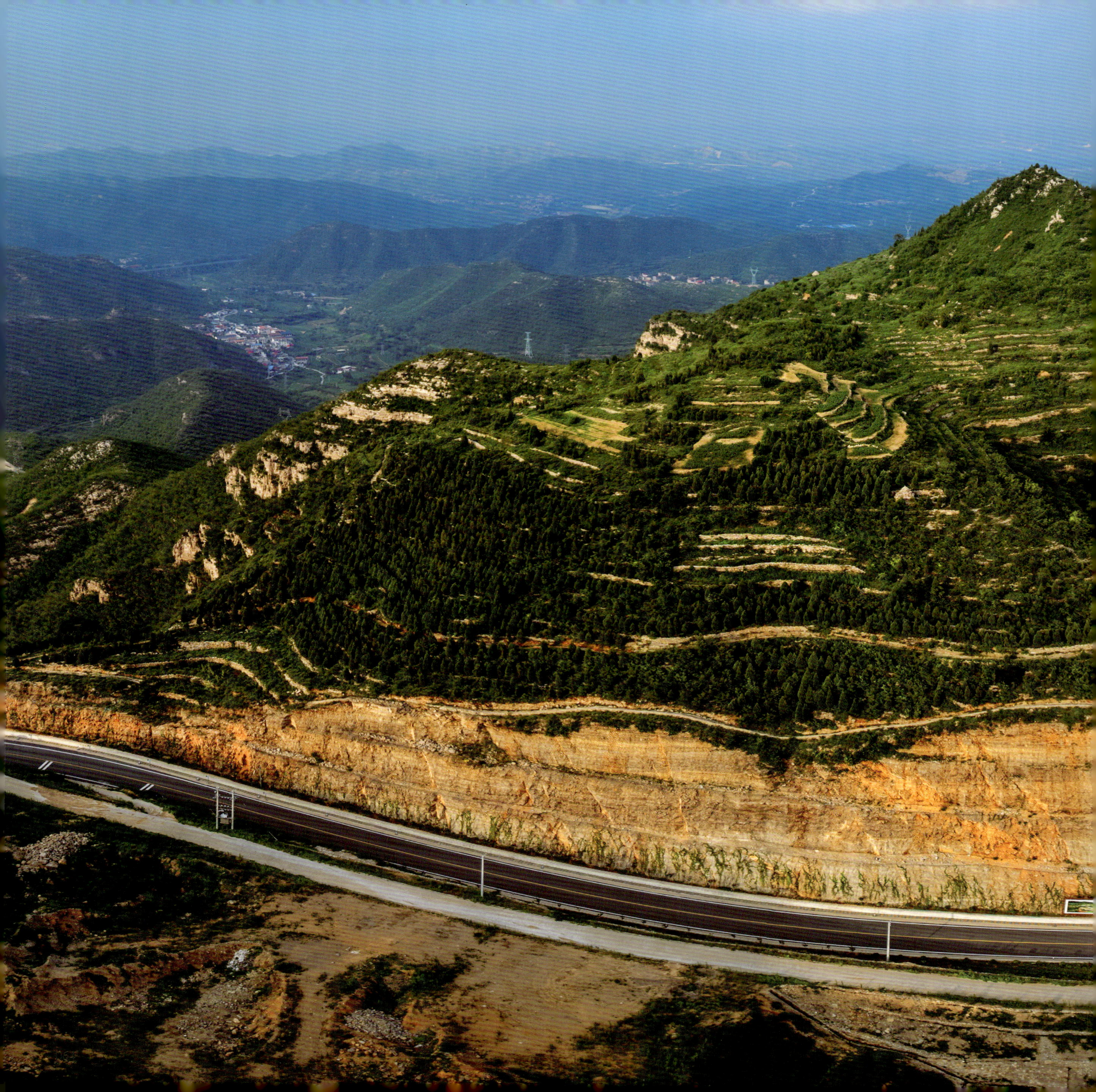

河北涉县旱作石堰梯田系统

邯郸市涉县形成了"叠石相次，包土成田"旱作石堰梯田系统，总面积约为178.67平方千米，被联合国世界粮食计划署专家称为"中国第二长城"，是全球重要农业文化遗产之一。

《河北涉县旱作石堰梯田》，邯郸市涉县，摄影：冯九伟

Dry Weir Terrace System in Shexian County, Hebei Province

Shexian County of Handan City formed a "stacked stones, covered with soil into fields" dry weir terrace system, with a total area approximately of 178.67 square kilometers, known by the United Nations World Food Programme experts as "the second Great Wall of China", is one of the world's important agricultural cultural heritage.

Dry Weir Terrace System in Shexian County, Hebei Province, Shexian County, Handan City Photographed by Feng Jiuwei

■ 河北涉县旱作石堰梯田系统

《涉县梯田》，邯郸市涉县，摄影：降国辉

Dry Weir Terrace System in Shexian County, Hebei Province
Terrace in Shexian County, Handan City, Shexian County, Handan City
Photographed by Jiang Guohui

河北涉县旱作石堰梯田系统

《人人有其田》，邯郸市涉县，摄影：郭静

Dry Weir Terrace System in Shexian County, Hebei Province
Everyone Has Their Own Land, Shexian County, Handan City
Photographed by Guo Jing

河北邯郸市邱县盘活低效用地"僵尸企业"

按照布局集中、用地集约、产业聚集的原则，邯郸高质量编制邯郸市国土空间规划，引导和促进产业项目向城镇、开发区集中，推动产业集聚高质量发展。当地政府开发政策措施，积极探索建立规划统筹、政府引导、市场运作、公众参与、利益共享的激励约束机制，充分利用存量土地资源发展新业态。此外，邯郸还按照"立足存量抓挖潜，节约集约促发展"的要求，加大存量土地盘活力度，加快推进土地资源"腾笼换鸟"，盘活闲置低效土地和"僵尸企业"，引导建设项目向园区集中，推动产业集约集聚发展，促进园区产业转型升级，让一些批而未供、供而未用的存量土地重新焕发生机。

《金色海岸》，邯郸市邱县，摄影：陈贵平

Revitalization of "Zombie Enterprises" with Low Land Use in Qiu County, Handan City, Hebei Province
In accordance with the principles of centralized layout, land use intensification and industrial agglomeration, high-quality creation of Handan City's territorial spatial planning has been carried out to guide and promote the concentration of industrial projects in towns and development zones, and to promote the high-quality development of industrial agglomeration. Local government committed to developing policies and measures, actively exploring the establishment of mechanism with planning, coordination, government guidance, market operation, public participation, benefit sharing incentives and constraints, and making full use of the land in stock to develop new business forms and patterns. In addition, Handan also in accordance with the requirements of "based on the land resource stock to grasp its potential, promotes economical and intensive development", promotes the land inventory revitalization, accelerates the implementation of "vacating cage to change birds" strategy in land resources development, revitalize inefficient land and "zombie enterprise", guides industrial projects to concentrate in parks, promotes the intensive and concentrated development of industry, promote industrial transformation and upgrading of the park, so that some of the land in stock which is granted but not in supply, or in supply but not for use is revitalized.

The Golden Coast, Qiu County, Handan City
Photographed by Chen Guiping

河北蔚县土地整治

河北蔚县1号沟土地整治后种植的油葵迎来大丰收。
《向日葵》，张家口市蔚县，摄影：马慧斌

Land Improvement in Yuxian County, Hebei Province
Sunflower planted after land reclamation in Ditch 1, Yuxian County, Hebei Province, has a bumper harvest.
Sunflower, Yuxian County, Zhangjiakou City
Photographed by Ma Huibin

矿产资源

八百里太行如绵延绿带，一座座名山如珍珠镶嵌其中；蕴藉深厚的燕山如同一道天然屏障，横亘在坝上高原与华北平原之间。这既成就了河北省地上的巍峨高峻，也造就了其地下的深厚底蕴。

矿产资源作为自然资源的重要组成部分，对经济社会发展起着重要支撑作用。河北省矿产资源丰富，主要分布于太行山和燕山山区。其中，煤炭主要分布于唐山、邯郸、邢台、张家口；铁矿主要分布于唐山、邯郸、邢台、承德；金、银等贵金属矿产主要分布于唐山、张家口、承德、石家庄；铅、锌、铜、钼等有色金属矿产主要分布于张家口、保定、承德；石灰岩、白云岩等建材类非金属矿产主要分布于浅山区；地热资源主要分布于河北平原中北部。

截至2022年，河北省已发现矿产166种，其中，有查明资源量的矿产139种。（河北省典型矿石种类见图3-1—图3-6）

河北省铁矿资源丰富，主要为火山型铁矿和沉积型铁矿。截至2022年，全省铁矿资源保有储量97.03亿吨，矿产地423处；铁矿企业585家，独立选矿企业2000余家，选矿能力远超采矿能力。

冀中煤炭基地是国家确定的13个煤炭基地之一，包括开滦、峰峰、邢台、井陉、蔚县、邯郸、宣化下花园、张家口北部8个大矿区和隆尧、大城平原含煤区。

河北省积极践行"两山"理念，不断优化矿山开发管理，构建矿产资源开发利用新格局。截至2022年，全省持有效采矿证的矿山数量有1766个，占全国的4.6%。

河北矿业持续向着绿色转型发展。截至2022年，全省67家矿山被纳入全国绿色矿山名录，数量居全国第四。

河北是全国率先开展大规模矿山生态修复的省份，这里涌现出三河东部矿区、武安响堂山、邯郸九龙山、秦皇岛栖云山等一大批矿山生态修复典型。

河北省地热资源比较丰富，分布广泛，且埋藏较浅，主要为对流型、传导型、多种成因叠加复合型地热资源等。根据地质构造及地形地貌，全省地热资源可划分为冀北山地区、冀西山地区、冀西北山间盆地、河北平原区四大热水区。已探明的深层地热资源和浅层地热能加在一起，可折算为1751.28亿吨标准煤。平原区地热水的可开采总量约为14.04亿立方米，实际年开采量约有5000多万立方米，占总量的3.6%。

Mineral Resources

Eight hundred miles Taihang Mountain looks like a stretching green belt, with famous mountains embedded in it like pearls; The rich Yanshan Mountain is a natural barrier between the Bashang Plateau and the North China Plain. This has not only achieved the lofty above ground of Hebei Province, but also created its deep underground heritage.

As one of the main elements of natural resources, mineral resources play an important supporting role in economic and social development. Hebei Province is rich in mineral resources, mainly distributed in the Taihang and Yanshan mountains. Among them, coal is mainly distributed in Tangshan, Handan, Xingtai and Zhangjiakou; iron ore is mainly distributed in Tangshan, Handan, Xingtai and Chengde; precious metals such as gold and silver are mainly distributed in Tangshan, Zhangjiakou, Chengde and Shijiazhuang; non-ferrous metals such as lead, zinc, copper and molybdenum are mainly distributed in Zhangjiakou, Baoding and Chengde. Non-metallic minerals of building materials such as limestone and dolomite are mainly distributed in shallow mountainous areas, and geothermal resources are mainly distributed in the central and northern Hebei Plain.

By 2022, 166 kinds of mineral resources have been discovered in Hebei Province, and 139 kinds of mineral resources have been identified. The pictures of typical ore types in Hebei Province are shown in Figure 3-1-Figure 3-6.

Hebei Province is rich in iron ore resources, mainly consisting of volcanic iron ore and sedimentary iron ore. As of 2022, the province's iron ore reserves are 9.703 billion tons, with 423 ore areas. There are 585 legitimate iron ore enterprises and more than 2,000 independent mineral processing enterprises, and the mineral processing capacity far exceeds the mining capacity.

Jizhong Coal Base is one of the 13 coal bases determined by the state, including 8 large mining areas in Kailuan, Fengfeng, Xingtai, Jingxing, Yuxian, Handan, Xuanhua Xiahuayuan, Zhangjiakou north and Longyao, Dacheng Plain coal bearing areas.

Hebei Province actively implements the concept of "two mountains", constantly optimizes mine development and management, and builds a new pattern of mineral resources development and utilization. By 2022, the number of mines holding valid mining certificates in the province is 1,766, accounting for 4.6% of the country.

Hebei mining industry continues to develop towards green transformation. By 2022, 67 mines in the province have been included in the national Green mine list, ranking fourth in the country.

As the first province in China to carry out large-scale mining ecological restoration, a large number of typical mining ecological restoration such as Sanhe East Mining Area, Wuan Xiangtang Mountain, Handan Jiulong Mountain, Qinhuangdao Qiyun Mountain and so on have emerged in Hebei.

The geothermal resources in Hebei Province are abundant, widely distributed and shallow buried, which are convection type, conduction type and superimposed compound of multiple causes. According to the regional geological structure and landform, the geothermal resources in the province can be divided into four hot water regions: North Mountain region, West Mountain region, northwest mountain basin and Hebei plain region. The proven deep geothermal resources and shallow geothermal energy combined can be converted to 175.128 billion tons of standard coal. In the plains, the total amount of geothermal water that can be extracted is about 1.404 billion cubic meters, while the actual annual extraction is more than about 50 million cubic meters, representing 3.6% of the total.

河北省典型矿石　TYPICAL ORES IN HEBEI PROVINCE

图 3-1 团块状铅锌银矿石　Figure 3-1 Massive lead-zinc-silver ore

图 3-2 角砾状铜钼矿石　Figure 3-2 Gravelly copper-molybdenum ore

图 3-3 磁铁矿　Figure 3-3 Magnetite

图 3-4 石灰岩　Figure 3-4 Limestone

图 3-5 赤铁矿　Figure 3-5 Hematite

图 3-6 金矿石　Figure 3-6 Gold ore

河北邯郸武安九龙山矿山生态修复公园

九龙山矿山生态修复公园位于邯郸武安市，由废弃矿山生态修复而成。这里山绿、水清、环境美。

《修复后的邯郸武安九龙山矿山生态公园全景》，邯郸市，摄影：宋现彬

Handan Wu'an Jiulongshan Mine Ecological Restoration Park
Jiulongshan Mine Ecological Restoration Park is located in Wu'an City, Handan. It is built from the ecological restoration of abandoned mines, with green mountains, clear water, and beautiful environment.
Panoramic view of the restored Jiulongshan Mine Ecological Park in Wu'an, Handan, Handan, Hebei Province, Handan City, Hebei Province
Photographed by Song Xianbin

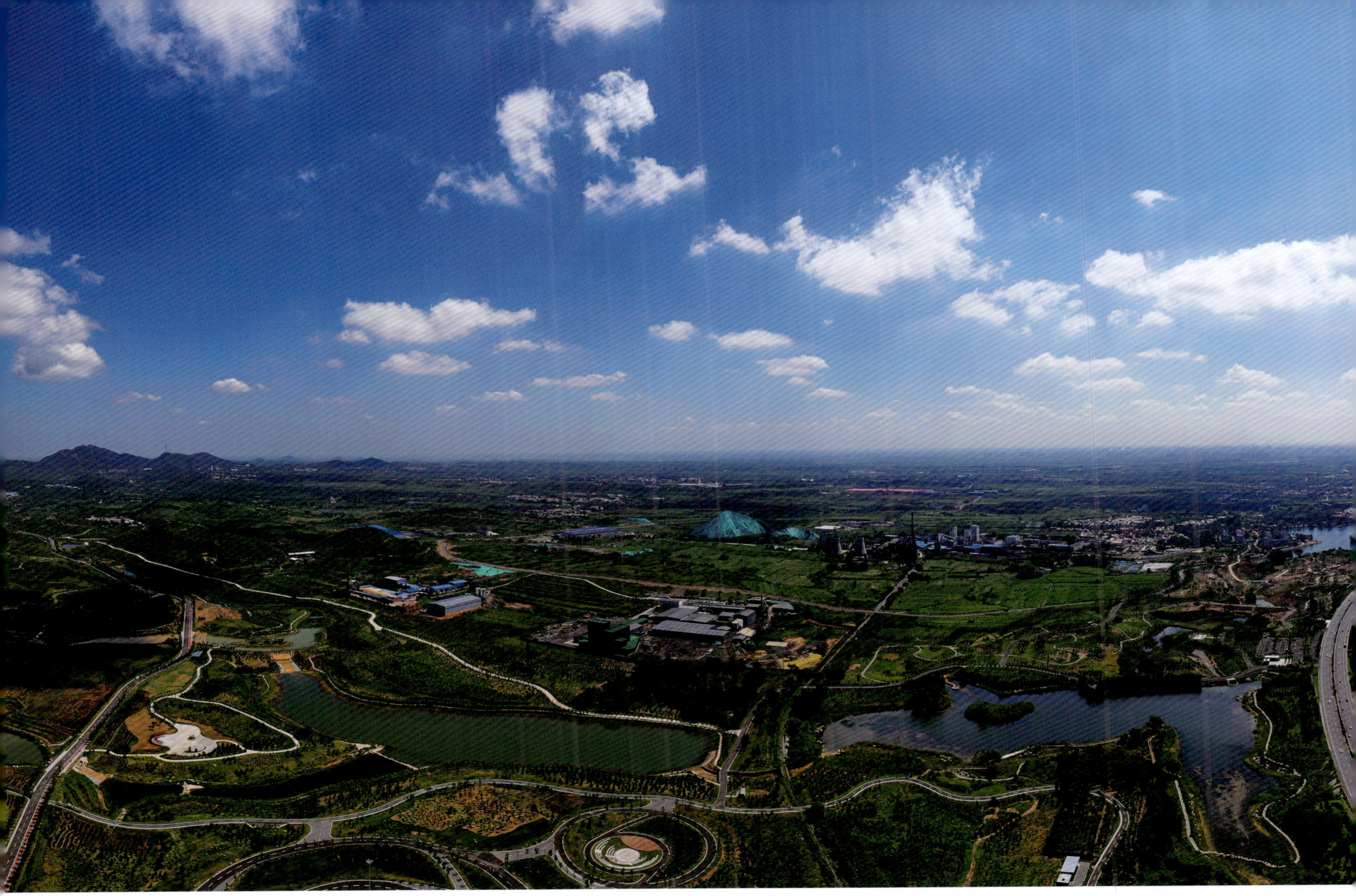

河北秦皇岛栖云山矿区

经过矿山修复治理，秦皇岛栖云山成为城市核心区的青山绿地公园。

《青山绿地公园》，秦皇岛市经开区栖云山，摄影：胡小龙

Qiyun Mountain Mining Area, Qinhuangdao, Heibei Province

Qiyun Mountain in Qinhuangdao has become a green park in the urban core after mine restoration and management.
Qingshan Greenland Park, Qiyun Mountain, Economic Development Zone, Qinhuangdao City
Photographed by Hu Xiaolong

河北三河市东部矿区

经过矿山修复治理三河市，东北部山区成为该市绿色发展的新引擎。
《五颜六色的格桑花竞相开放》三河市东北部山区，摄影：李连明

Eastern Mining Area of Sanhe City, Hebei Province
The mountainous area in northeastern Sanhe City has become a new engine for the city's green development after mine restoration and treatment.
Colorful Gesang flowers are blooming, Mountainous area northeast of Sanhe City
Photographed by Li Lianming

河北宽城满族自治县东梁金矿区

通过实施矿山环境治理修复工程,东梁金矿满山覆绿,环境优美。
《绿色生态的东梁金矿》,承德市宽城县,摄影:章洪涛

Dongliang Gold Mining Area, Kuancheng Manchu Autonomous County, Hebei Province
After mine restoration and management, Dongliang Gold Mine is covered in green mountains and has a beautiful environment.
Green and Ecological Dongliang Gold Mine, Kuancheng Manchu Autonomous County, Chengde City
Photographed by Zhang Hongtao

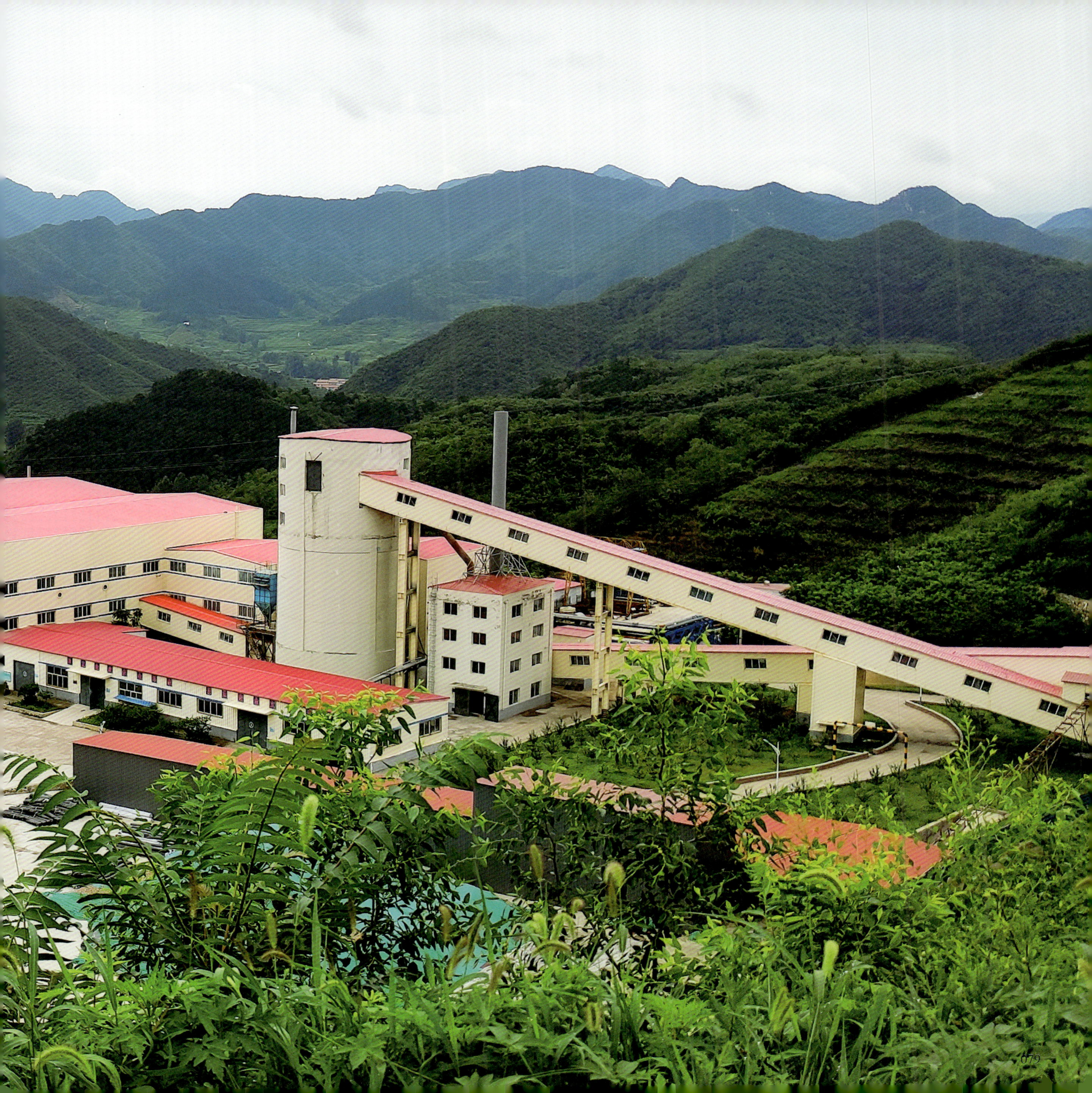

森林资源

森林,是生命的殿堂、万物的摇篮。树木的根系深入土壤,能防止水土流失。森林中的空气清新湿润,充满负氧离子。参天大树用茂密的枝叶编织出生命的密网,动物穿梭其间,构成和谐而神秘的生态世界。

截至2021年,河北省森林面积约为4.6052万平方千米,森林覆盖率24.41%,活立木蓄积约为19041.72万立方米,森林蓄积约15101.10万立方米。

河北省木兰围场国有林场始建于1963年,总经营面积约为1058.40平方千米。这里地处内蒙古浑善达克沙地南缘的滦河上游地区、承德市围场县境内,是京津冀地区重要水源涵养地和防风阻沙的重要生态屏障。截至2022年,该林场林地面积由建场初的200平方千米增至约897.53平方千米,活立木蓄积由62万立方米增至约814.8万立方米,森林覆盖率由15.5%增至85.5%,林业建设取得显著成绩。

塞罕坝地处河北最北部、内蒙古浑善达克沙地南缘,属典型的半干旱半湿润寒温性大陆季风气候区。

塞罕坝机械林场总经营面积约933.33平方千米,其中林地面积约767.33平方千米,林木蓄积量约为1036.8万立方米,森林覆盖率82%,成为京津地区的"水源卫士"、生态屏障。

2021年8月,习近平总书记在承德考察时强调:"塞罕坝林场建设史是一部可歌可泣的艰苦奋斗史。你们用实际行动铸就了牢记使命、艰苦创业、绿色发展的塞罕坝精神,这对全国生态文明建设具有重要示范意义。"

雾灵山位于河北省承德市兴隆县境内,属暖温带落叶阔叶林带向温带针阔混交林带的过渡地带,具有明显的森林植被垂直分布带特征。

雾灵山生物多样性十分丰富,有高等植物1871种,其中国家二级保护植物15种,雾灵丁香、雾灵景天等植物37种。陆生脊椎动物248种,其中国家一级保护动物6种、二级保护动物35种。大型真菌238种、昆虫3200余种。

Forest Resources

The forest is the temple of life and the cradle of all things. The root system of trees penetrates deep into the soil and prevents soil erosion. The air in the forest is fresh and moist, full of negative oxygen ions. Towering trees weave a dense web of life with dense branches and leaves, and animals shuttle between them, forming a harmonious and mysterious ecological world.

As of 2021, Hebei Province has a forest area of approximately 46,052 square kilometers, a forest coverage rate of 24.41%, a standing tree stock of approximately 190,417,200 cubic meters, and a forest stock of approximately 151,011,000 cubic meters.

Founded in 1963 with a total operating area of approximately 1,058.40 Square kilometers, Hebei Mulan Paddock State-owned Forest Farm is located in the upper reaches of the Luanhe River on the southern edge of the Hunshandak Sandy Land in Inner Mongolia and in Weichang County, Chengde City, which is an important water conservation area and an important ecological barrier for wind and sand control in the Beijing-Tianjin-Hebei region. As of 2022, the forest land area of the forest farm has increased from approximately 200 Square kilometers at the beginning of the establishment to approximately 897.53 Square kilometers, the stock of standing trees has increased from 620,000 cubic meters to approximately 8,148,000 cubic meters, and the forest coverage rate has increased from 15.5% to 85.5%.

Saihanba is located in the northernmost part of Hebei Province and the southern edge of the Hunshandak Sandy Land in Inner Mongolia, which is a typical semi-arid and semi-humid cold temperate continental monsoon climate zone.

Saihanba Machinery Forest Farm has a total operating area of approximately 933.33 Square kilometers, a forest land area of approximately 767.33 Square kilometers, a forest stock of about 10.368 million cubic meters, and a forest coverage rate of 82%, becoming a "water source guard" and ecological barrier in the Beijing-Tianjin area.

In August 2021, General Secretary Xi Jinping emphasized during his inspection in Chengde: "The history of the construction of Saihanba Forest Farm is a history of hard work that can be sung and cried. You have forged the Saihanba spirit of keeping in mind the mission, hard work and green development with practical actions, which is of important demonstration significance for the construction of ecological civilization in the country."

Wuling Mountain is located in Xinglong County, Chengde City, Hebei Province, which belongs to the transition zone from warm temperate deciduous broad-leaved forest belt to temperate coniferous and broad-leaved mixed forest belt, with obvious characteristics of vertical distribution zone of forest vegetation.

Wuling Mountain is very rich in biodiversity, with 1,871 species of higher plants, including 15 species of national second-class protected plants; 37 species of model plants such as Lilac and Sedum sedum; there are 248 species of terrestrial vertebrates, including 6 species of national first-class protected animals and 35 species of second-class protected animals; there are 238 species of macrofungi and more than 3,200 species of insects.

河北木兰围场国有林场

河北省木兰围场国有林场位于河北省承德市围场满族蒙古族自治县境内，主要由塞罕坝国家森林公园、御道口草原森林风景区和红松洼国家级自然保护区等三大景区组成，是京津冀重要的水源涵养功能区和生态安全屏障。相关数据见表4-1。
《木兰围场》，承德市围场县，摄影：刘云华

Hebei Mulan Paddock State-owned Forest Farm
Located within the territory of Weichang Manchu and Mongolian Autonomous County in Chengde City, Hebei Province, it mainly consists of three major scenic areas, including Saihanba National Forest Park, Yudaokou Grassland and Forest Scenic Area, and Hongsongwa National Nature Reserve. It is an important water conservation functional area and ecological security barrier for Beijing, Tianjin and Hebei.
Mulan Paddock, Weichang Manchu and Mongolian Autonomous County, Chengde City
Photographed by Liu Yunhua

表 4-1 河北省木兰围场国有林场相关数据（2022年底）
Table 4-1 Related data of Mulan Paddock State-owned Forest Farm (end of 2022)

总经营面积 Total operating area	1058.40 平方千米 1058.40 square kilometers
林地面积 Area of woodland	897.53 平方千米 897.53 square kilometers
活立木蓄积 Accumulation of standing trees	814.8 万立方米 8,148,000 cubic meters
森林覆盖率 Forest cover	85.5%

表4-2 雾灵山保护区森林植被垂直分布带
Table 4-2 Vertical distribution zone of forest vegetation in Wuling Mountain National Nature Reserve

海拔 Elevation	分布带 Distribution bands	主要树种 Major tree species
1900米以上 1900 meters or more	亚高山草甸灌丛带 Subalpine meadow shrub belt	
1700—1900米之间 Between 1700 and 1900 meters	针叶林带 Coniferous forest belt	落叶松 larch
1600—1700米之间 Between 1600 and 1700 meters	针阔混交林带 Mixed coniferous and broad-leaved forest belt	
1000—1600米之间 Between 1000 and 1600 meters	落叶阔叶林带 Deciduous broad-leaved forest belt	桦树林、杨桦混交林 Birch forest, poplar and birch mixed forest
1000米以下 Below 1000 meters	松砾林带 Pine gravel forest belt	

河北雾灵山国家级自然保护区

雾灵山为国家级自然保护区，森林覆盖率高达93%，属温带大陆季风性山地气候，具有雨热同季、冬长夏短、四季分明、夏季凉爽、昼夜温差大的特征。保护区森林植被垂直分布情况见表4-2。

《雾灵山》，承德市兴隆县，摄影：李连明

Hebei Wuling Mountain National Nature Reserve

As a national nature reserve, it has a high forest coverage rate of 93%. It features a temperate continental monsoon mountain climate, characterized by concurrent rainy and hot seasons, long winters and short summers, distinct four seasons, cool summers, and significant temperature differences between day and night.

Wuling Mountain, Xinglong County, Chengde City
Photographed by Li Lianming

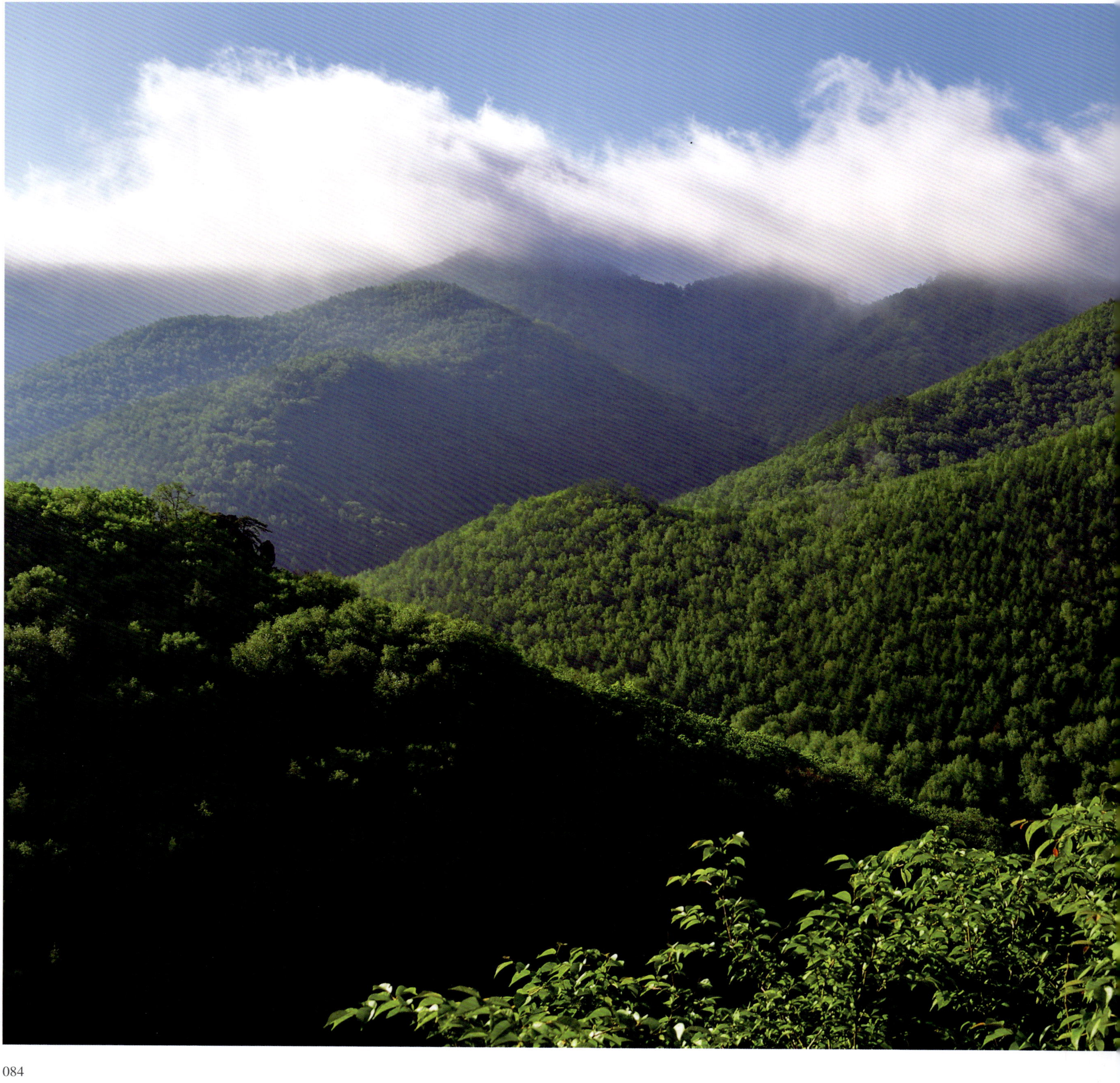

表 4-3 塞罕坝全年气候概况表
Table 4-3 Annual Climate Overview of Saihanba

海拔 Elevation	年均气温 Average annual temperature	积雪期 Snow period	无霜期 Frost-free period	降水量 precipitation	大风日数 Number of windy days	气候区 Climatic zone
1010-1939.9 米 1010-1939.9m	零下 1.3°C -1.3°C	7 个月 7 Months	64 天 64 Days	460 毫米 460mm	53 天 53 days	半干旱半湿润寒温性大陆季风气候区 Semi-arid and semi-humid cold temperate continental monsoon climate zone

河北塞罕坝机械林场

　　塞罕坝机械林场位于河北省承德市围场满族蒙古族自治县北部坝上地区，是清朝时期著名的皇家猎苑"木兰围场"的重要组成部分。其境内是滦河、辽河的发源地之一。塞罕坝全年气候概况见表 4-3。
《塞罕坝机械林场阴河分场》，承德市围场县，摄影：袁春龙

Saihanba Mechanical Forest Farm, Hebei Province

Located in the northern Bashang area of Weichang Manchu and Mongolian Autonomous County, Chengde City, Hebei Province, it is an important part of the famous Qing Dynasty royal hunting ground "Mulan Paddock". The territory is one of the source places of Luanhe River and Liaohe River.
Yinhe Branch of Saihanba Mechanical Forest Farm, Weichang Manchu and Mongolian Autonomous County, Chengde City
Photographed by Yuan Chunlong

河北塞罕坝机械林场

《塞罕坝林海》（组图），承德市围场县，摄影：林树国

Saihanba Mechanical Forest Farm, Hebei Province
The Forest Sea of Saihanba Mechanical Forest Farm (Series), Weichang Manchu and Mongolian Autonomous County, Chengde City
Photographed by Lin Shuguo

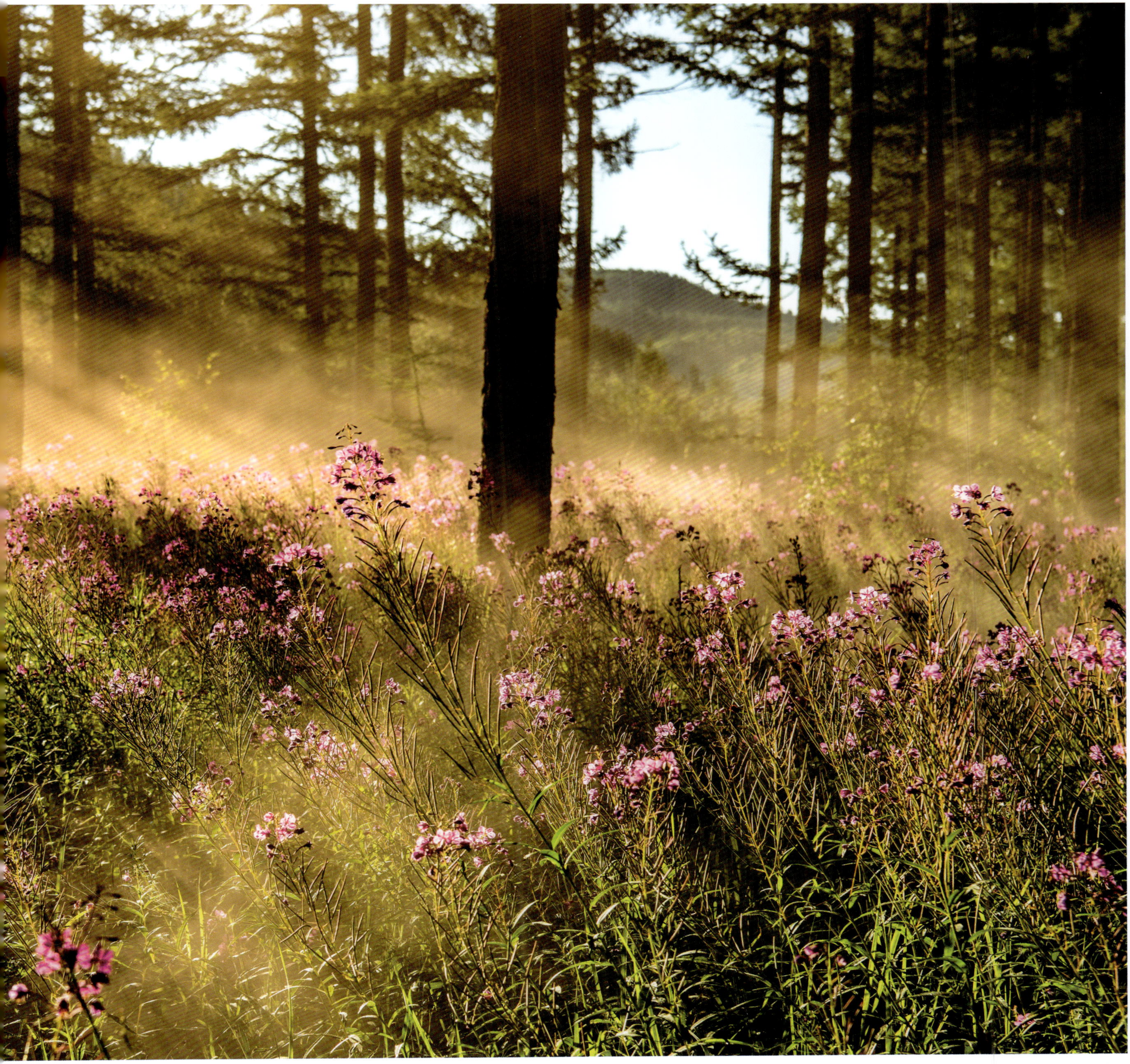

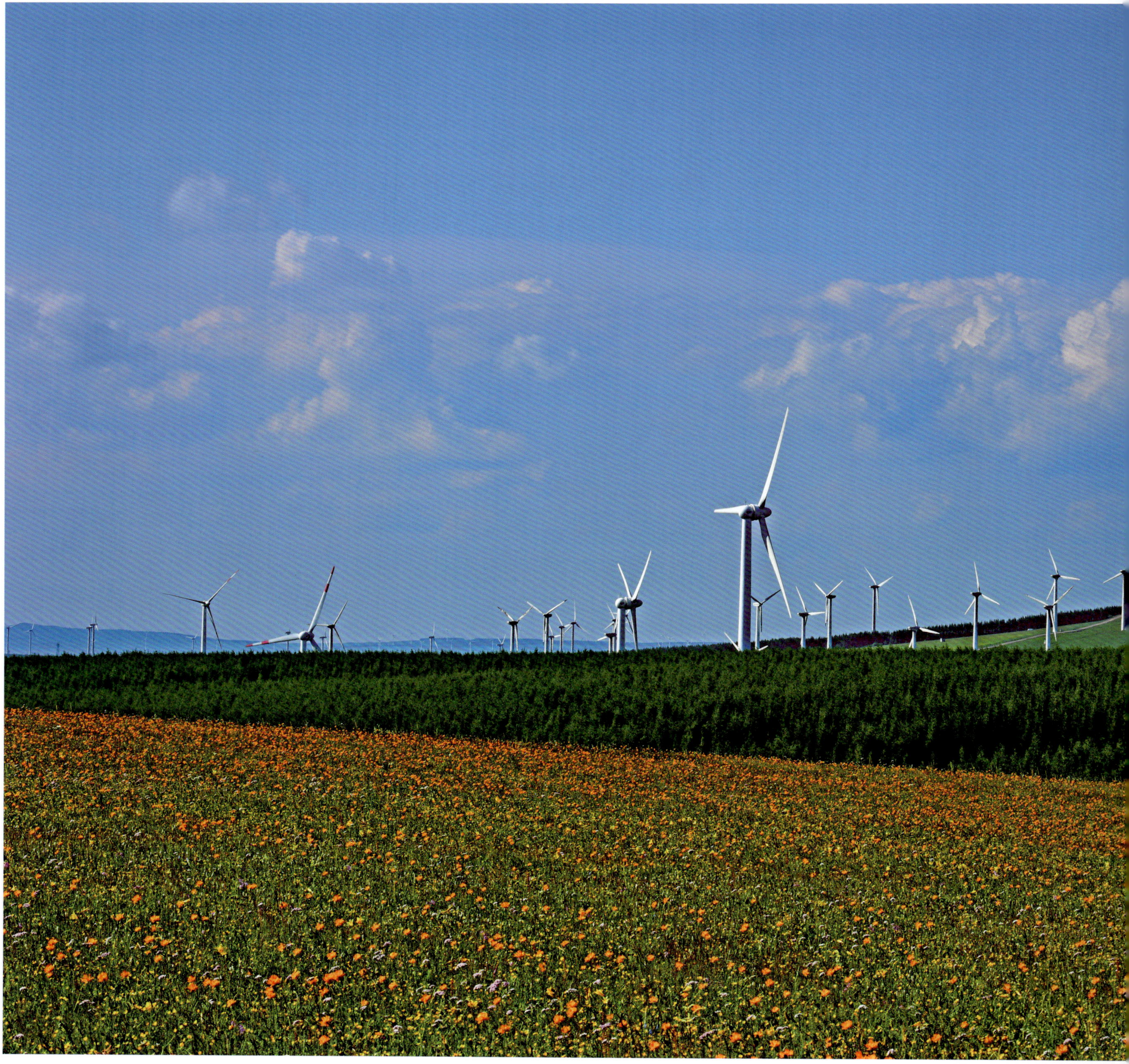

■ 河北塞罕坝机械林场

《旱地金莲》（左图），承德市围场县围场红松洼自然保护区
《七星湖》（右上图），承德市围场县塞罕坝千层板分场七星湖
《泰丰湖日出》（右下图），承德市围场县塞罕坝千层板分场
摄影：林树国

Hebei Saihanba Mechanical Forest Farm (Chengde City, Weichang County)
Dryland Nasturtium (left), Weichang Hongsongwa Nature Reserve
Seven Star Lake (Upper Right), Saihanba mille-feuille panel is divided into Seven Star Lake
Sunrise over Taifeng Lake (Below Right), a branch of the mille-feuille plate of Saihanba
Photographed by Lin Shuguo

河北塞罕坝机械林场

《林海中的守望》（上），承德市围场县，摄影：林树国
《塞罕坝机械林场》（下），承德市围场县，韩宏亮
《塞罕坝机械林场》（右），承德市围场县，劳志刚

Saihanba Mechanical Forest Farm, Hebei Province
The Watch in the Sea of Forests (Above), Weichang County, Chengde City, Photographed by Lin Shuguo
Saihanba Mechanical Forest Farm (Bottom), Weichang County, Chengde City Photographed by Han Hongliang
Saihanba Mechanical Forest Farm (Right), Weichang County, Chengde City Photographed by Lao Zhigang

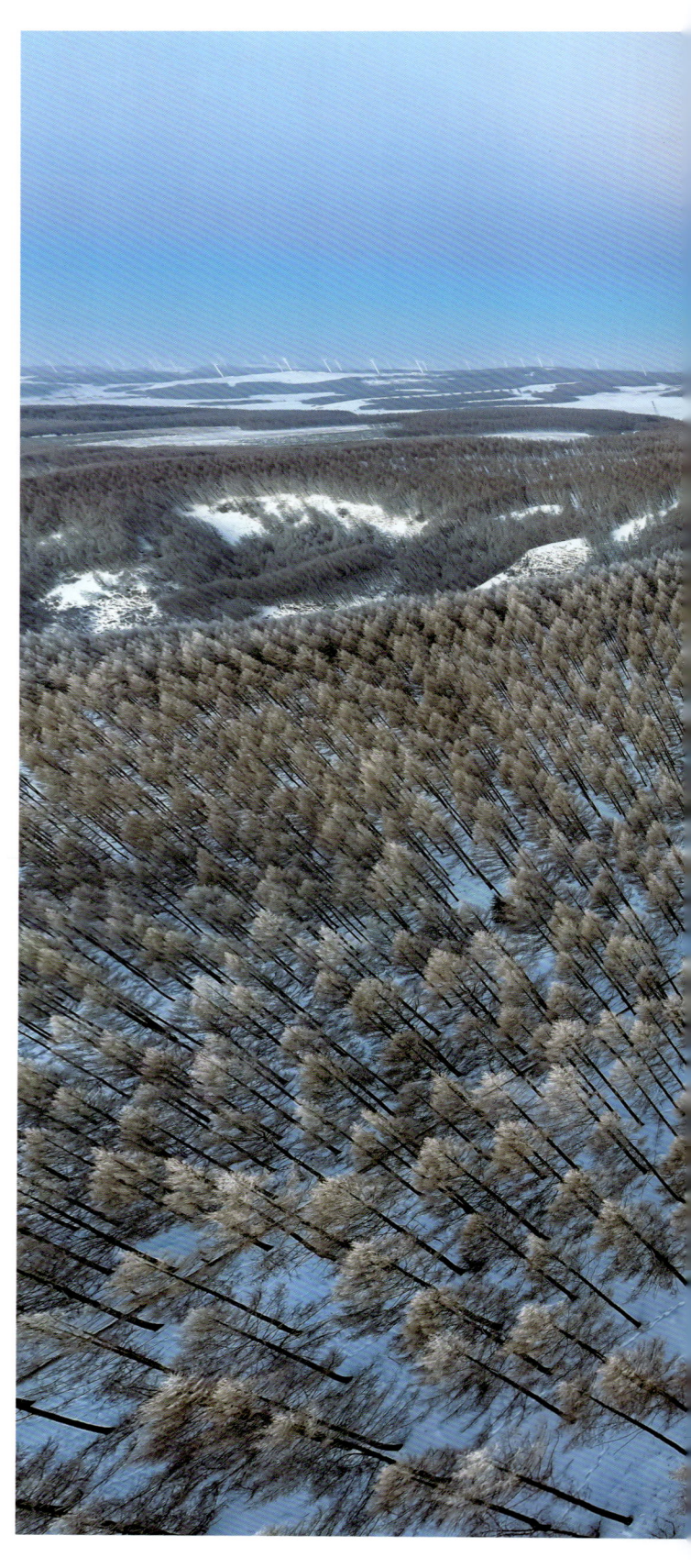

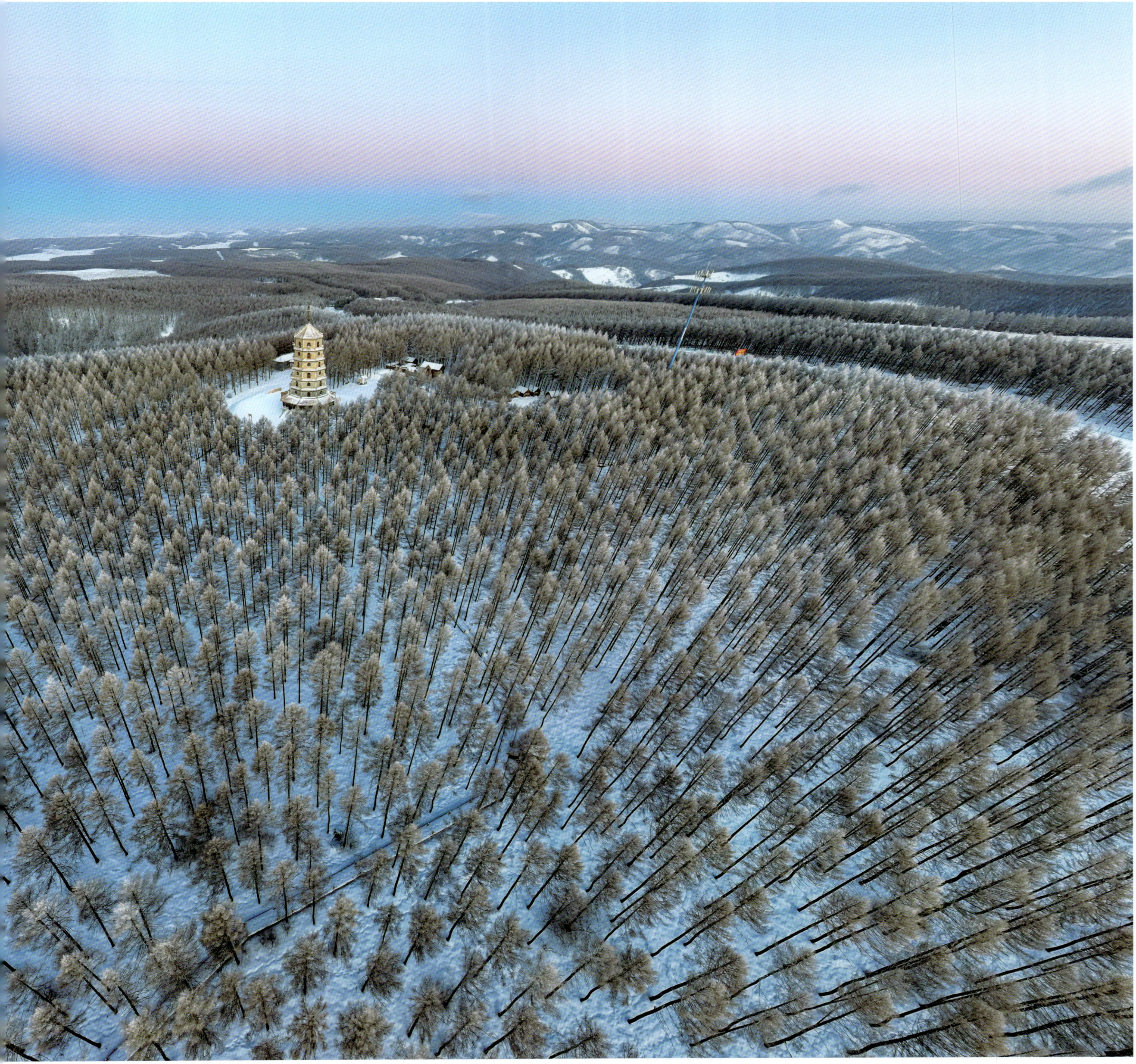

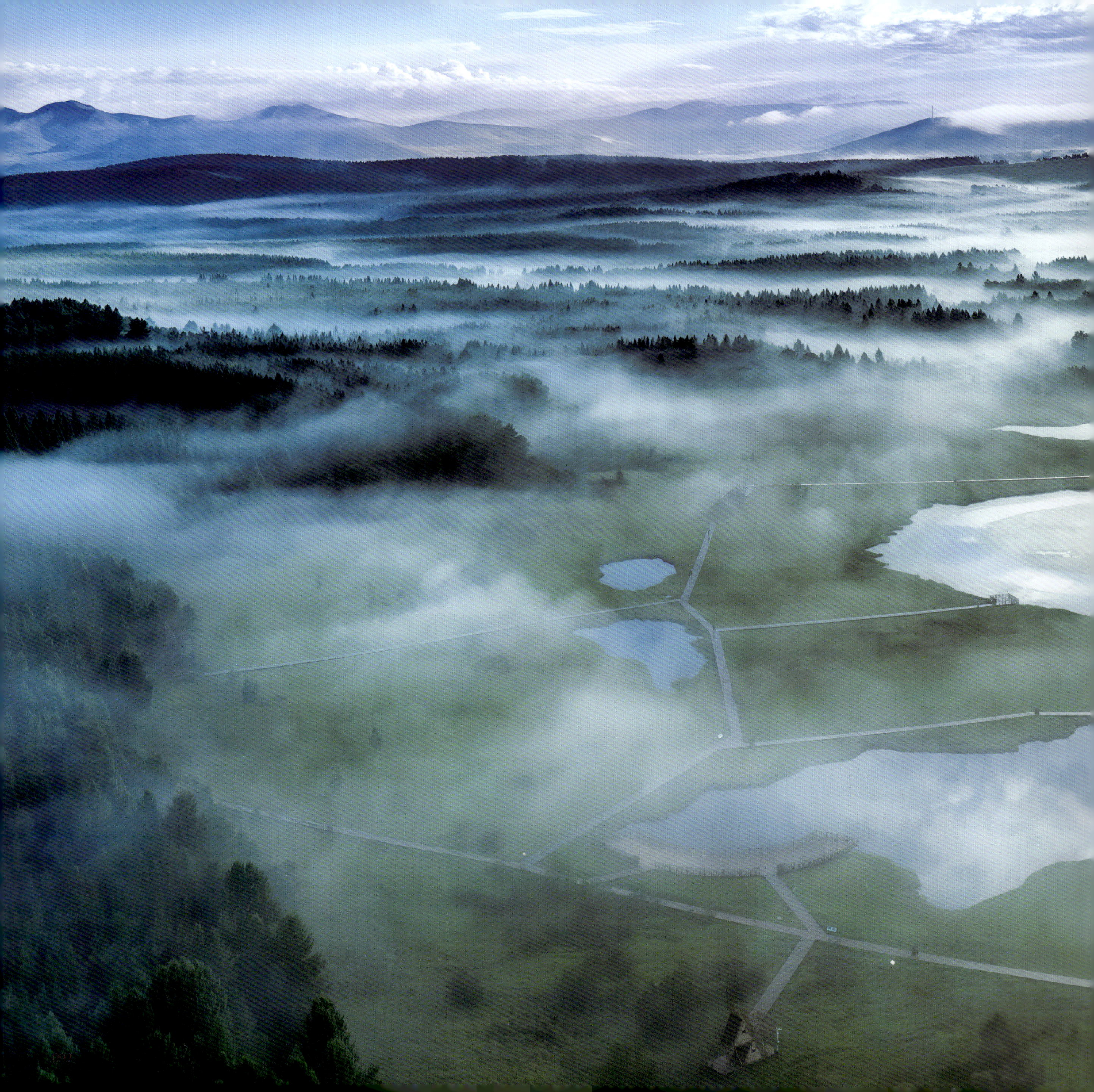

■ 河北塞罕坝机械林场

《承德塞罕坝七星湖湿地》，承德市围场县，摄影：周刚

Saihanba Mechanical Forest Farm, Hebei Province
The Chengde Saihanba Seven-Star Lake Wetland, Weichang
Manchu and Mongolian Autonomous County, Chengde City
Photographed by Zhou Gang

草原资源

草原，是大地的绿色诗篇。成群的牛羊和骏马在其中自由驰骋，尽情展现生命的力与美。野花在草丛中静静绽放，以微小的生命点缀着草原的辽阔。

河北省是我国北方的重要草原区，草原是其绿色优势、生态屏障、资源禀赋。根据2022年国土变更调查数据，全省草原面积约为19472.67平方千米，其中，天然牧草地约有4199.9平方千米、人工牧草地约有109.67平方千米、其他草地约有15163.2平方千米。河北各市均有草地分布，主要分布于张家口、承德、保定三市，占全省草原总面积的84%。

河北黄土湾国家草原自然公园是首批国家草原自然公园试点之一。公园位于张家口市塞北管理区黄土湾村。该公园以"黄金草原、生态画廊"为形象定位，以保护温性典型草原生态系统与科学利用草原资源为主要目标，开展生态保护修复、生态旅游、科研监测及自然宣教等活动的特定区域。

蔚县空中草原位于蔚县古城南15公里处，地处山巅，海拔约为2158米，面积约36平方千米。这里日均气温为15℃左右，多种气候条件并存，天气瞬息万变，夏季无酷暑，春季无沙尘，空气清新。

草原天路位于张家口市北部，地处内蒙古高原与华北平原结合处、阴山山脉与燕山山脉交汇处，横跨沽源、崇礼、张北、万全、尚义五区县，是河北省"十四五"旅游产业发展的重中之重。草原天路全长约为323.9千米，沿途既有历史古迹，也有地质景观，还有草原美景。

京北第一草原位于丰宁县西北部的大滩镇，距承德市区约250公里、距北京市约280公里，是离首都最近的天然草原。该草原地处高原坝上，平均海拔1487米，总面积约350平方千米，动植物资源十分丰富。夏季，这里最高气温不超24℃，是避暑的理想之所。

Grassland Resources

The grassland is the green poetry of the earth. Herds of cattle, sheep and horses gallop freely in it, showing the power and beauty of life to the fullest. Wildflowers bloom quietly in the grass, dotting the vast expanse of the grassland with tiny life.

Hebei Province is an important grassland area in northern China, and grassland is its green advantage, ecological barrier, and resource endowment. According the 2022 land change survey data, the province's grassland area is approximately 19,472.67 square kilometers. Among them, about 4,199.8 square kilometers of natural pasture land, about 109.67 square kilometers of artificial pasture land, and about 15,163.2 square kilometers of other grassland. There are grasslands in all cities of Hebei province, mainly in Zhangjiakou, Chengde and Baoding, accounting for 84% of the total grassland area of the province.

Hebei Huangtuwan National Grassland Nature Park is one of the first national grassland nature park pilots. The park is located in Huangtuwan Village, Saibei Management District, Zhangjiakou City. The park is positioned as a "golden grassland and ecological gallery", and is a specific area for ecological protection and restoration, eco-tourism, scientific research monitoring and nature education with the main goal of protecting the temperate typical grassland ecosystem and scientifically utilizing grassland resources.

The Sky Grassland of Yu County is located 15 kilometers south of the ancient city of Yuxian County, on the top of the mountain, with an altitude of about 2,158 meters and an area of about 36 square kilometers. The average daily temperature here is about 15 °C, a variety of climatic conditions coexist, the weather is changing rapidly, there is no heat in summer, no sand and dust in spring, and the air is fresh.

Located in the north of Zhangjiakou City, at the junction of the Inner Mongolia Plateau and the North China Plain, at the intersection of the Yinshan Mountains and the Yanshan Mountains, across the five districts and counties of Guyuan, Chongli, Zhangbei, Wanquan and Shangyi, Grassland Sky Road is the top priority of the development of the tourism industry in Hebei Province during the 14th Five-Year Plan. The total length is about 323.9 kilometers, along the way there are historical monuments, geological landscapes and grassland beauty.

Located in Datan Town, northwest of Fengning County, about 250 kilometers away from Chengde City and about 280 kilometers away from Beijing, The First Grassland of Northern Beijing is the closest natural grassland to the capital. The grassland is located on the plateau dam, with an average altitude of 1,487 meters and a total area of approximately 350 square kilometers, which is very rich in animal and plant resources. The maximum temperature in summer does not exceed 24°C, making it an ideal place to escape the heat.

河北黄土湾国家草原自然公园

　　河北黄土湾国家草原自然公园位于河北省张家口市塞北管理区黄土湾村，是国家林业和草原局公布的首批国家草原自然公园试点。该公园以"黄金草原、生态画廊"为形象定位，是以保护温性典型草原生态系统与科学利用草原资源为主要目标，开展生态保护修复、生态旅游、科研监测及自然宣教等活动的特定区域。
《河北黄土湾国家草原自然公园》，张家口市塞北管理区，摄影：方四成

Huangtuwan National Grassland Nature Park , Hebei Province
Located in Huangtuwan Village, Saibei Management District, Zhangjiakou City, Hebei Province, it is one of the first national grassland nature park pilots announced by the State Forestry and Grassland Administration. The park is positioned as a "golden grassland, ecological gallery", with its primary goals being the protection of temperate typical grassland ecosystems and the scientific utilization of grassland resources. It serves as a designated area for ecological protection and restoration, eco-tourism, scientific research monitoring, and natural education activities.
Hebei Huangtuwan National Grassland Nature Park, Saibei Management District, Zhangjiakou City
Photographed by Fang Sicheng

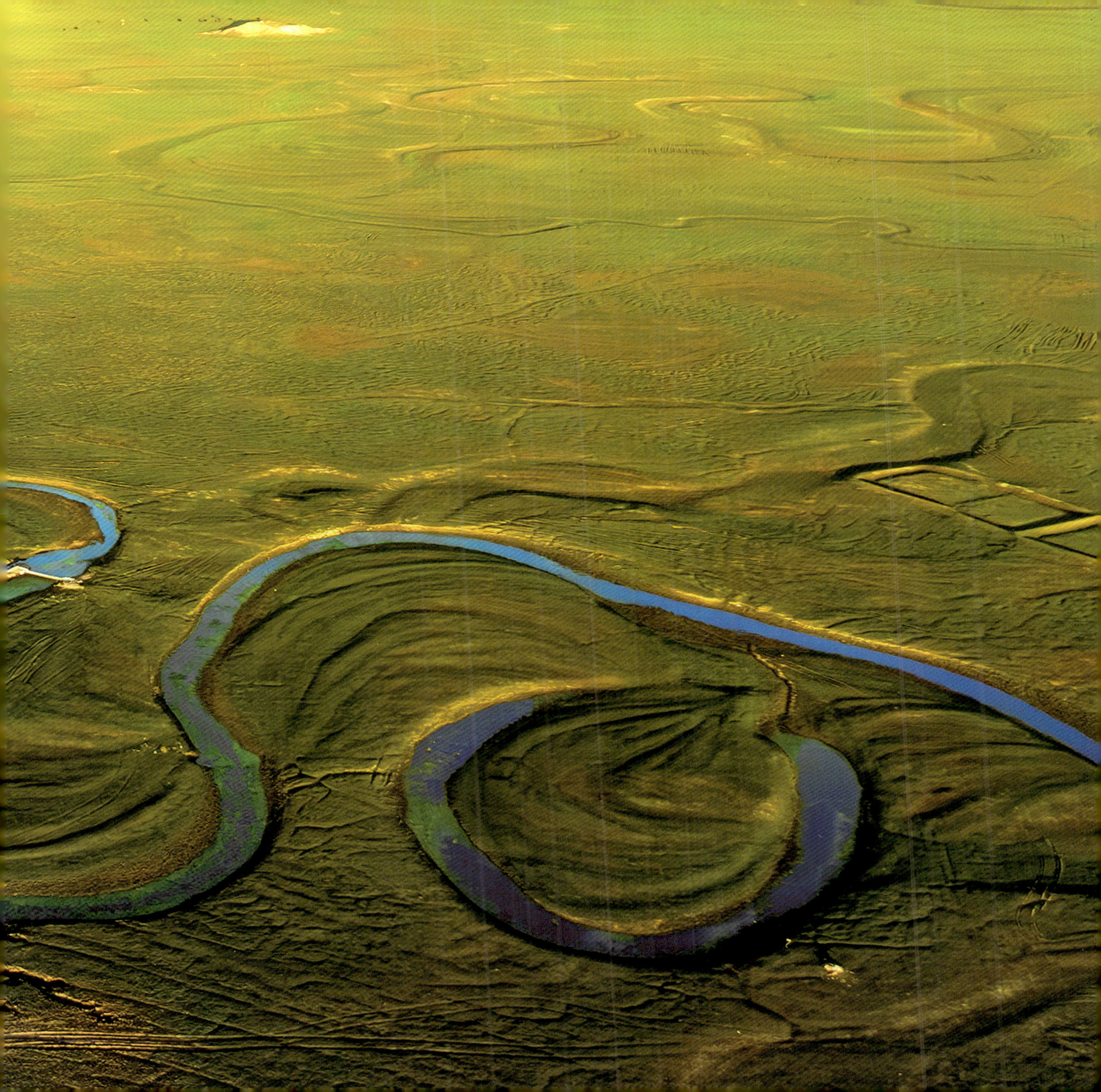

《草原晴空白云飘》，张家口市张北县，摄影：孙苓阁

Grassland under a clear sky with white clouds drifting, Zhangbei County, Zhangjiakou City
Photography by Sun Lingge

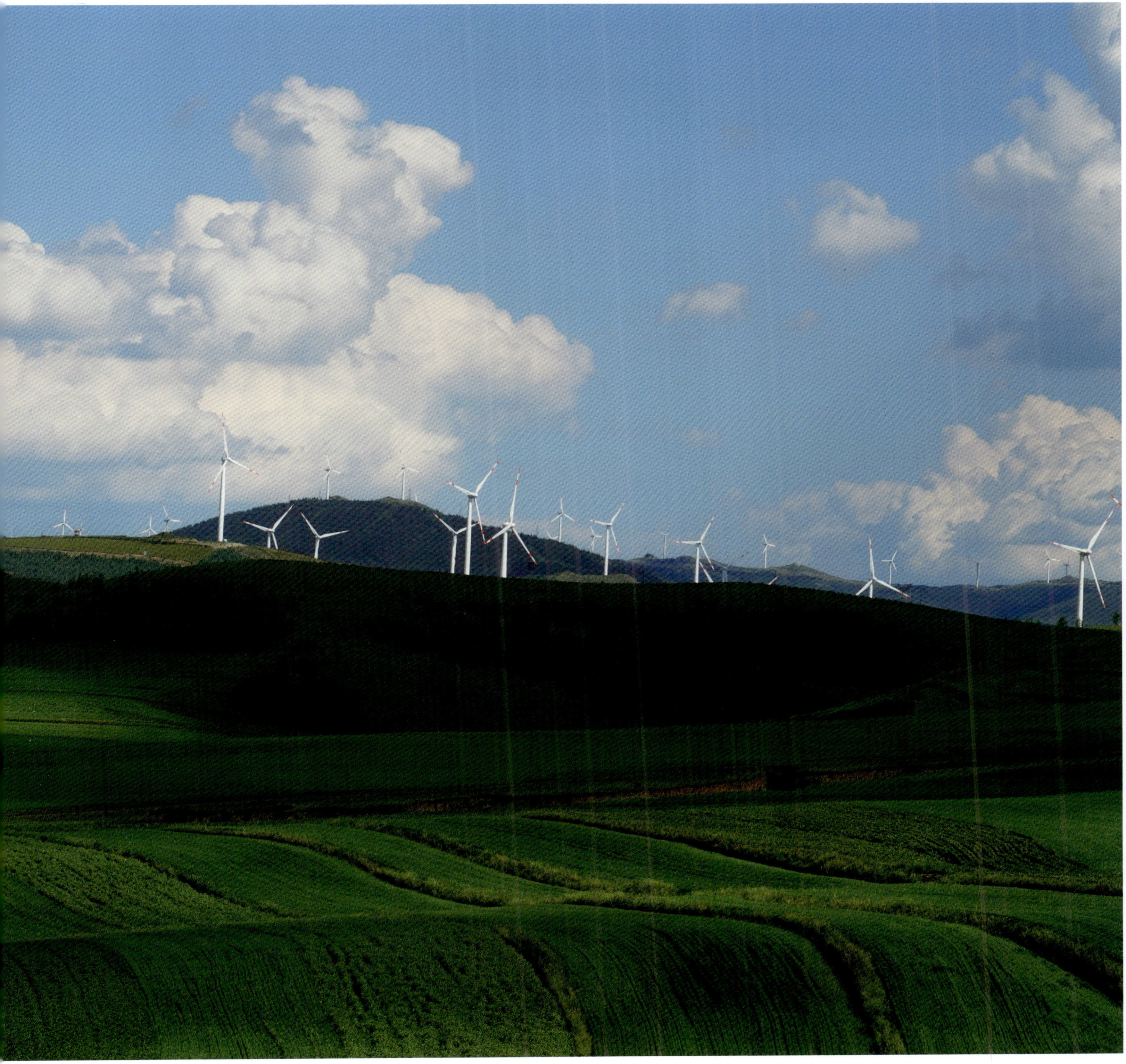

河北蔚县空中草原

蔚县空中草原位于河北省蔚县。这里属暖温带季风气候，四季分明。
《蔚县空中草原》，张家口市蔚县，摄影：刘洋

Yuxian Sky Grassland, Hebei Province
Situated in Yuxian County, Hebei Province, the Yuxian Sky Grassland boasts a warm temperate monsoon climate characterized by four distinct seasons.
Yuxian Sky Grassland, Yuxian County, Zhangjiakou City
Photography by Liu Yang

河北京北第一草原

　　京北第一草原位于河北省承德市丰宁满族自治县大滩镇。这里以其广阔的草原、清凉的夏季和丰富的旅游活动而闻名。

《京北第一草原》，承德市丰宁县，供图：视觉中国

The First Grassland of Northern Beijing, Hebei Province

Located in Datan Town, Fengning Manchu Autonomous County, Chengde City, Hebei Province, The First Grassland of Northern Beijing is famous for its vast grassland, cool summer and rich tourism activities.

The First Grassland of Northern Beijing, Fengning Manchu Autonomous County, Chengde City
Photo provided by Visual China

河北张家口草原天路

张家口草原天路位于张家口市张北县和崇礼区的交界处，是连接崇礼滑雪区、赤城温泉区、张北草原风景区、白龙洞风景区、大青山风景区的一条重要通道，也是"中国大陆"十大最美丽公路之一。
《林海长龙》，张家口市张北县和崇礼区的交界处，摄影：袁晓虹

Zhangjiakou Grassland Sky Road, Hebei Province
Situated at the junction of Zhangbei County and Chongli District in Zhangjiakou City, this road acts as an important passage connecting Chongli Ski Area, Chicheng Hot Spring Area, Zhangbei Grassland Scenic Area, Bailong Cave Scenic Area and Daqingshan Scenic Area. Furthermore, it ranks among mainland China's top ten most scenic roads.
The Road Winds Like a Long Dragon Throuth the Forest Sea, The junction of Zhangbei County and Chongli District, Zhangjiakou City
Photography Yuan Xiaohong

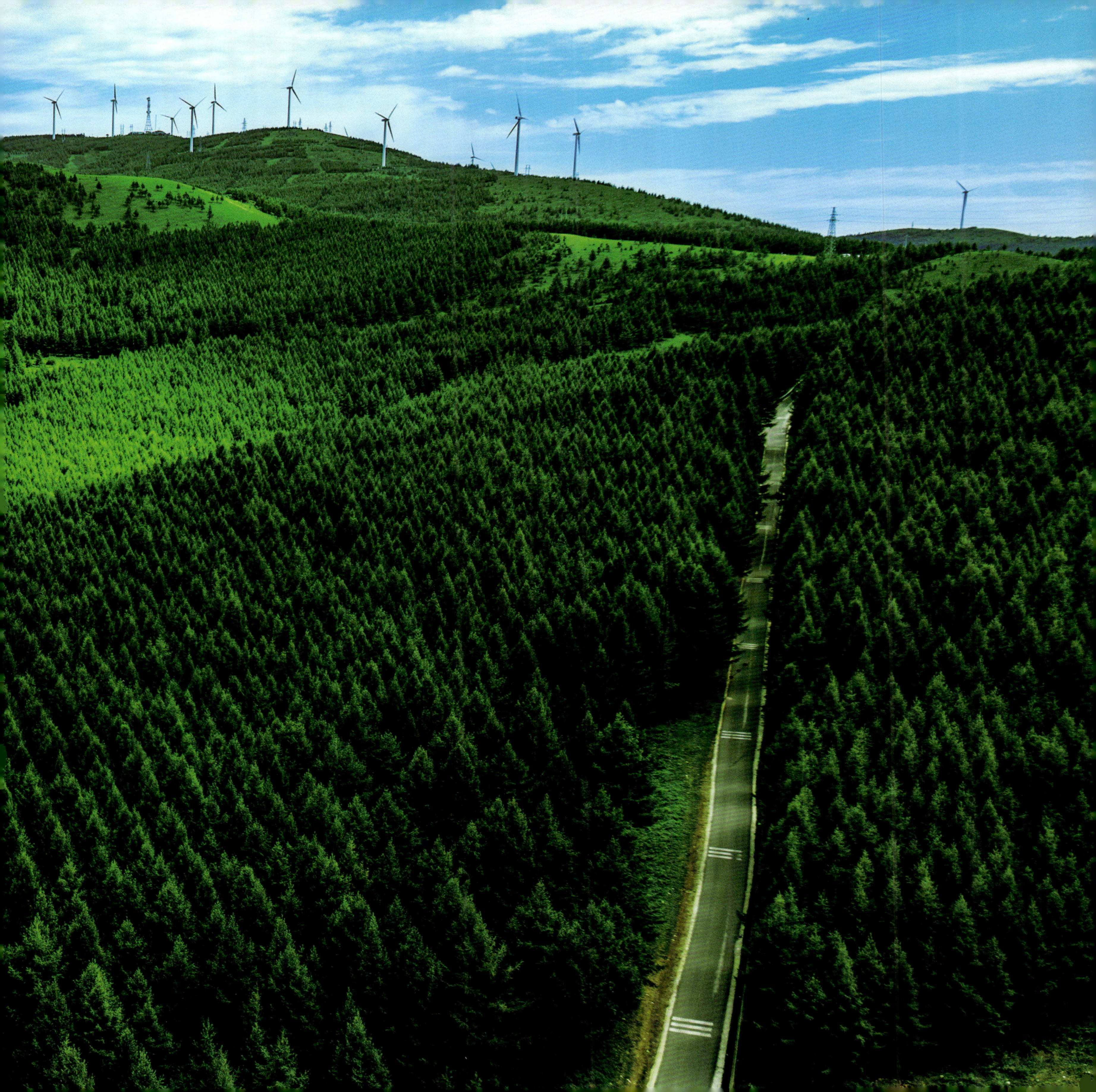

湿地资源

湿地，是大自然恩赐的绿洲。芦苇摇曳、飞鸟展翅、鱼儿畅游，还有默默生长的植物，构成了和谐的生态世界。湿地像海绵一样吸收雨水，减缓了洪水的冲击；又像过滤器一样净化水质，让清澈的水流源源不断滋养大地。

湿地是重要的生态系统，具有涵养水源、调节气候、改善环境、维护生物多样性等生态功能，与人类的生存发展息息相关，被誉为"地球之肾"。

河北省兼有高原、山地、丘陵、平原、湖泊、海滨等地貌类型，孕育了丰富的湿地资源。河北省湿地资源丰富，类型多样，包括森林沼泽、灌丛沼泽、沼泽草地、沿海滩涂、内陆滩涂、沼泽地、河流水面、湖泊水面、水库水面、坑塘水面、沟渠和浅海水域12个湿地类型。根据河北省2022年度国土变更调查数据和河北省海洋基础测绘成果统计，河北省湿地总面积7268万平方千米。

闪电河国家湿地公园位于河北省沽源县城东部，距县城5公里。这里是河流、沼泽、湖泊、积水洼地等组合而成的复合型湿地，也是鸟类南北、东西迁徙的交汇地和多种珍稀濒危鸟类的迁徙停歇地。

闪电河国家湿地公园的动植物资源丰富，约有植物297种，陆生野生脊椎动物274种，占全省陆生脊椎动物总量的41.2%。其中，鸟类多达222种，较2019年增加35种。

怀来官厅水库国家湿地公园位于张家口市怀来县城南部，主要分布于永定河和官厅水库消落带区域，东西横跨怀来县中部，规划总面积约为135.33平方千米。官厅水库入库水质由Ⅳ类提升至Ⅲ类，湿地公园野生植物由106种增至318种，野生鸟类由169种增至192种。

南大港湿地位于沧州市南大港产业园区东北部，是国家重要湿地、全省首个自然资源确权登记试点。沧州南大港候鸟栖息地是河北省第一个世界自然遗产地。

南大港湿地有众多珍稀濒危鸟类及其他野生动植物繁衍生息，其中包括国家一级保护鸟类东方白鹳、黑脸琵鹭、卷羽鹈鹕。该湿地2023年共监测记录鸟类10万余只，特别是在环境修复区域，鸟类的种类、数量及珍稀程度均明显提升。

Wetland Resources

Wetlands are oases of nature's gifts. The swaying reeds, the birds spreading their wings, the fish swimming, and the plants that grow silently constitute a harmonious ecological world. Wetlands act like sponges to absorb rainwater and slow down the impact of flooding; It also purifies the water like a filter, allowing the clear water to continue to nourish the earth.

Wetlands are important ecosystems with ecological functions such as water conservation, climate regulation, environmental improvement, and biodiversity maintenance, which are closely related to the survival and development of human beings, and are known as the "kidneys of the earth".

Hebei Province has both plateau, mountain, hill, plain, lake, seashore and other landform types, giving birth to rich wetland resources. Hebei Province is rich in wetland resources, including forest swamp, shrub swamp, swamp grassland, along the beach, inland tidal flat, swamp, river water surface, lake water surface, reservoir water surface, pit pond water surface, ditch and shallow sea water 12 wetland types. According to the 2022 land change survey data of Hebei Province and the statistics of Hebei Province's marine basic surveying and mapping results, the total wetland area of Hebei province is 7,268 square kilometers.

Located in the east of Guyuan County, Hebei Province, 5 kilometers away from the county seat, Lightning River National Wetland Park is a composite wetland composed of rivers, swamps, lakes and stagnant depressions, and is the intersection of north-south and east-west migrations of birds and the migration and stopping place of a variety of rare and endangered birds.

The Lightning River National Wetland Park is rich in animal and plant resources, with 297 species of plants and 274 species of terrestrial wild vertebrates, accounting for 41.2% of the total terrestrial vertebrates in the province. Among them, there are as many as 222 species of birds, an increase of 35 species from 2019.

Huailai Guanting Reservoir National Wetland Park is located in the south of Huailai County, Zhangjiakou City, mainly distributed in the Yongding River and Guanting Reservoir subsidence zone, across the central part of Huailai County from east to west, with a total planned area of about 135.33 square kilometers. The water quality of Guanting Reservoir has been improved from Class IV to Class III, the number of wild plant species in the wetland park has increased from 106 to 318, and the number of wild bird species has increased from 169 to 192.

Nandagang Wetland is located in the northeast of Nandagang Industrial Park, Cangzhou City. Here is a wetland of national importance and the first pilot project for the confirmation and registration of natural resource rights in the province. Cangzhou Nandagang Migratory Bird Habitat is the first World Natural Heritage Site in Hebei province.

Nandagang Wetland is home to many rare and endangered birds and other wild animals and plants, including the Oriental White Stork, Black-faced Spoonbill and Curly-feathered Pelican, which all are national first-class protected birds. In 2023, more than 100,000 birds will be monitored and recorded in the wetland, especially in the restoration area, the species, number and rarity of birds have increased significantly.

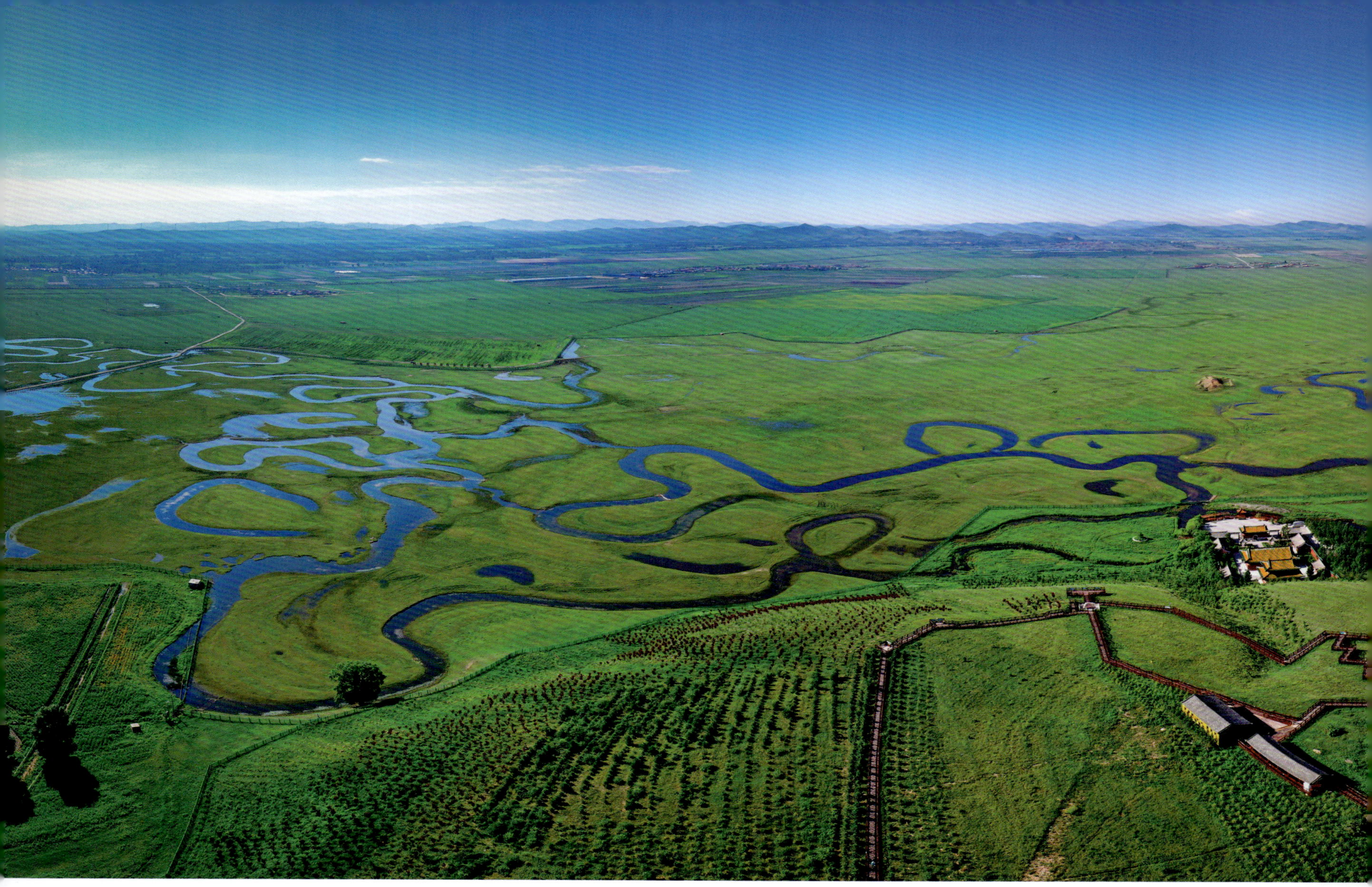

河北坝上闪电河国家湿地公园

　　河北坝上闪电河国家湿地公园位于河北省张家口市沽源县城东部。这里是由河流、湖泊、滩涂、沼泽和沼泽化草甸以及人工库塘组成的复合型内陆湿地。

《滦河神韵全景》，张家口市沽源县，摄影：李颂

Bashang Lightning River National Wetland Park, Hebei Province

Located in the eastern part of Guyuan County, Zhangjiakou City, Hebei Province, the Bashang Lightning River National Wetland Park is a complex inland wetland consisting of rivers, lakes, mudflats, swamps, swampy meadows and artificial reservoirs.

Luanhe Charm Panoramic View, Guyuan County, Zhangjiakou City
Photographed by Li Song

河北陆生脊椎动物总量 Total terrestrial vertebrates in Hebei	占比 Percentage
闪电河国家湿地公园 Lightning River National Wetland Park	41.2%
其他 Other	58.8%

河北坝上闪电河国家湿地公园

《闪电河国家湿地公园》，张家口市沽源县，摄影：王爱忠

Bashang Lightning River National Wetland Park, Hebei Province

Lightning River National Wetland Park in Hebei province, Guyuan County, Zhangjiakou City
Photographed by Wang Aizhong

■ 河北坝上闪电河国家湿地公园

《龙凤呈祥》，张家口市沽源县，摄影：任峻封

Bashang Lightning River National Wetland Park, Hebei Province

Auspiciousness Presented by Dragon and Phoenix, Guyuan County, Zhangjiakou City
Photographed by Ren Junfeng

河北怀来官厅水库国家湿地公园

　　河北怀来官厅水库国家湿地公园位于张家口市怀来县城区南部，以保护怀来县官厅水库及上游永定河区域湿地生态环境和野生动植物为主。这里的湿地类型包括水库水面、内陆滩涂、坑塘水面（非养殖）、沼泽草地、河流水面和沟渠等6种类型。
《官厅水库国家湿地公园航拍》，张家口市怀来县，摄影：尤志婷

Huailai Guanting Reservoir National Wetland Park, Hebei Province
Located in the southern part of Huailai County, Zhangjiakou City, its main focus is on protecting the wetland ecological environment and wildlife of the Guanting Reservoir and the upstream area of the Yongding River. There are six types of wetlands here, including reservoir surface, inland mudflats, pond surface (non-aquaculture), marsh grassland, river surface and ditches.
Aerial Photography of Guanting Reservoir National Wetland Park, Huailai County, Zhangjiakou City
Photographed by You Zhiting

河北北戴河国家湿地公园

　　北戴河国家湿地公园位于秦皇岛市北戴河区中部沿海区域，是依托北戴河湿地大潮坪和新河水系及沿海防护林形成的湿地生态系统。这里主要有近海与海岸湿地、河流湿地、沼泽湿地及人工湿地等多种湿地类型。这里独特的地理位置和丰富的湿地及森林资源成为多种珍稀鸟类栖息的天堂和城市生态环境的绿色屏障。

《珍珠海滩》，秦皇岛市北戴河区，摄影：周秦明

Beidaihe National Wetland Park, Hebei Province

Located in the central coastal area of Beidaihe District, Qinhuangdao City, it is a wetland ecosystem formed by the Beidaihe wetland tidal flat, the Xinhe River system and the coastal shelterbelt. There are many types of wetlands, including offshore and coastal wetlands, river wetlands, swamp wetlands and artificial wetlands. Its unique geographical location and rich wetland and forest resources have become a paradise for many rare birds and a green barrier for the urban ecological environment.

Pearl Beach, Beidaihe District, Qinhuangdao City
Photographed by Zhou Qinming

《北戴河国家湿地公园》（一），秦皇岛市北戴河区，摄影：田利君

Beidaihe National Wetland Park (1), Beidaihe District, Qinhuangdao City
Photographed by Tian Lijun

《北戴河国家湿地公园》（二），秦皇岛市北戴河区，摄影：田利君

Beidaihe National Wetland Park (2), Beidaihe District, Qinhuangdao City
Photographed by Tian Lijun

河北南大港湿地

　　南大港湿地位于河北省沧州市东北部，是著名的退海河流淤积型滨海湿地，由草甸、沼泽、水体、野生动植物等多种生态要素组成。南大港候鸟栖息地成为河北省第一个世界自然遗产地。
《湿地"青蛙图"》，沧州市东北部，摄影：张洪彬

Nandagang Wetland, Hebei Province
Located in the northeastern part of Cangzhou City, Hebei Province, Nandagang Wetland is a famous coastal wetland of river sedimentation type, consisting of meadows, swamps, water bodies, wild animals and plants and other ecological elements. The Nandagang migratory bird habitat has become the first world natural heritage site in Hebei province.
Wetland Frog Photograph, northeastern part of Cangzhou City
Photographed by Zhang Hongbin

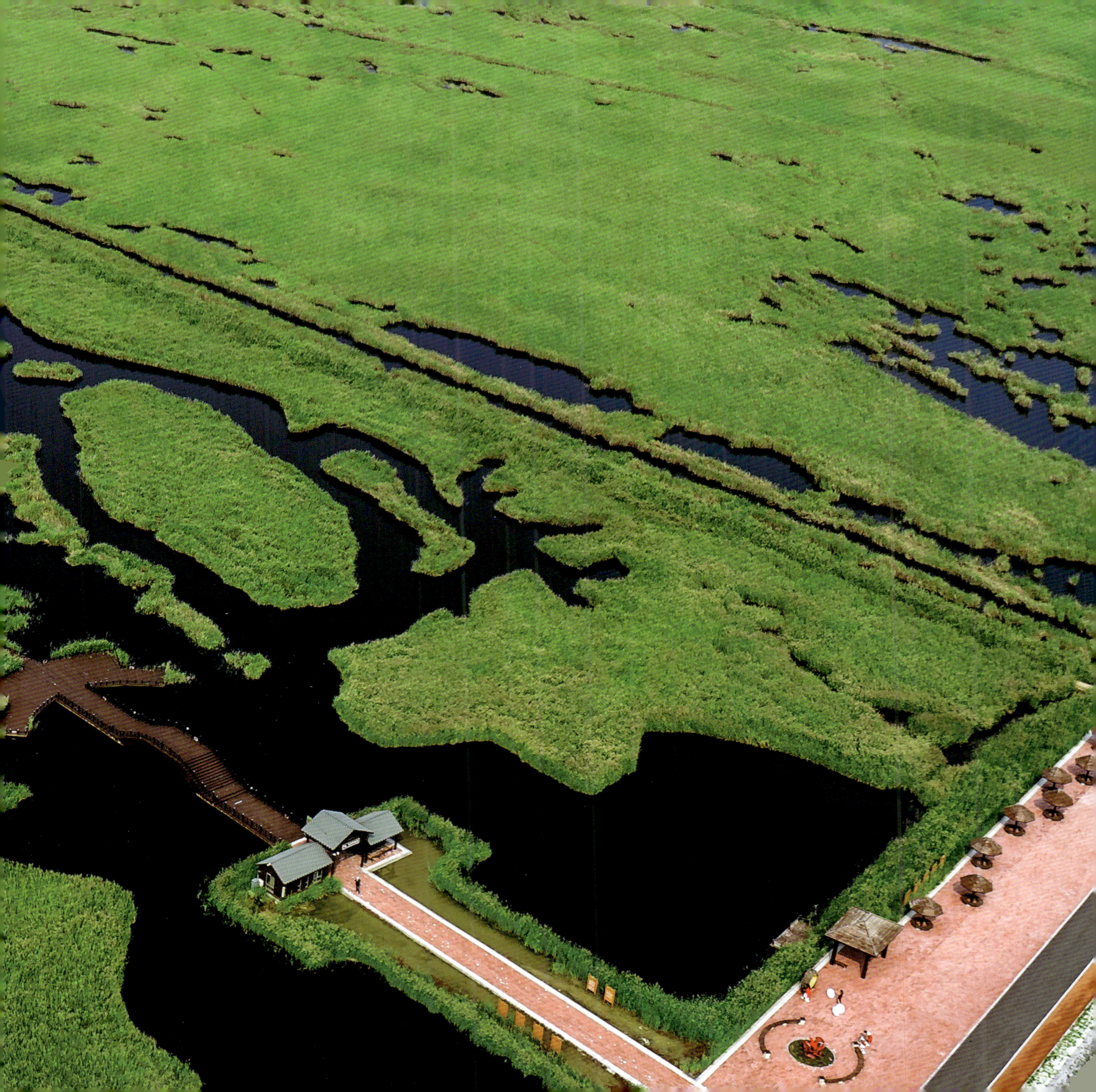

水资源

河北省境内地势起伏较大，受地形、降水等因素影响，这里发育了不同水系。这里河流众多，呈扇形分布。河北省的水系由三部分组成：海河流域、辽河流域（含大凌河）和内蒙古高原内流区。其中，海河流域水系由漳卫南运河、子牙河、大清河、永定河、北运河、潮白河、蓟运河等组成，占全省总面积的66.4%。

根据河北省第一次水利普查成果，河北省流域面积200—3000平方千米的中小河流共有264条，河流在省内总长度约为13600多千米。

河北省水资源地区分布很不平衡，山前区水资源条件好，黑龙港和沿海地区水资源缺乏。地表水主要分布在山区，占地表水总量的79%。平原河川径流量绝大多数由汛期几次降水形成，非汛期多数河道断流。

2023年，河北全省水资源总量约为241.35亿立方米，比上一年增加53.38亿立方米，比多年平均值多了64.88亿立方米。其中，地表水资源量约为121.85亿立方米，地下水资源量约为182.75亿立方米；人均及亩均水资源量分别为326立方米和267立方米。各设区市中，保定市水资源总量最多，约有57.57亿立方米，占全省水资源总量的23.9%。

Water Resources

The terrain in Hebei Province is very undulating, affected by terrain, precipitation and other factors, different water systems have developed here, with numerous rivers distributed in a fan-shaped pattern. The river system of Hebei Province consists of three parts: the Haihe River basin, the Liaohe River basin (including the Daling River), and the Inner flow area of Mongolia Plateau. Among them, the Haihe River system is composed of Zhangwei South Canal, Ziya River, Daqing River, Yongding River, North Canal, Chaobai River and Ji Canal, accounting for 66.4% of the total area of the province.

According to the results of the first water conservancy survey in Hebei Province, there are 264 small and medium-sized rivers with a drainage area of 200 to 3,000 square kilometers in Hebei province, with a total length of more than 13,600 kilometers in the province.

The regional distribution of water resources in Hebei Province is very unbalanced, the water resources in the piedmont area are good, and the water resources in Heilong Port and coastal areas are short. Surface water is mainly distributed in mountainous areas, accounting for 79% of the total surface water. Most of the river runoff in the plain is formed by several precipitation in flood season, and most of the river channels are cut off in non-flood season.

In 2023, the total water resources of Hebei Province will be about 24.135 billion cubic meters, an increase of 5.338 billion cubic meters over the previous year and 6.488 billion cubic meters more than the multi-year average. Among them, the surface water resources are about 12.185 billion cubic meters, the groundwater resources are about 18.275 billion cubic meters, and the per capita and per mu water resources are 326 cubic meters and 267 cubic meters, respectively. Among the administrative units Baoding City has the largest total water resources, with about 5.757 billion cubic meters, accounting for 23.9% of the total water resources of the province.

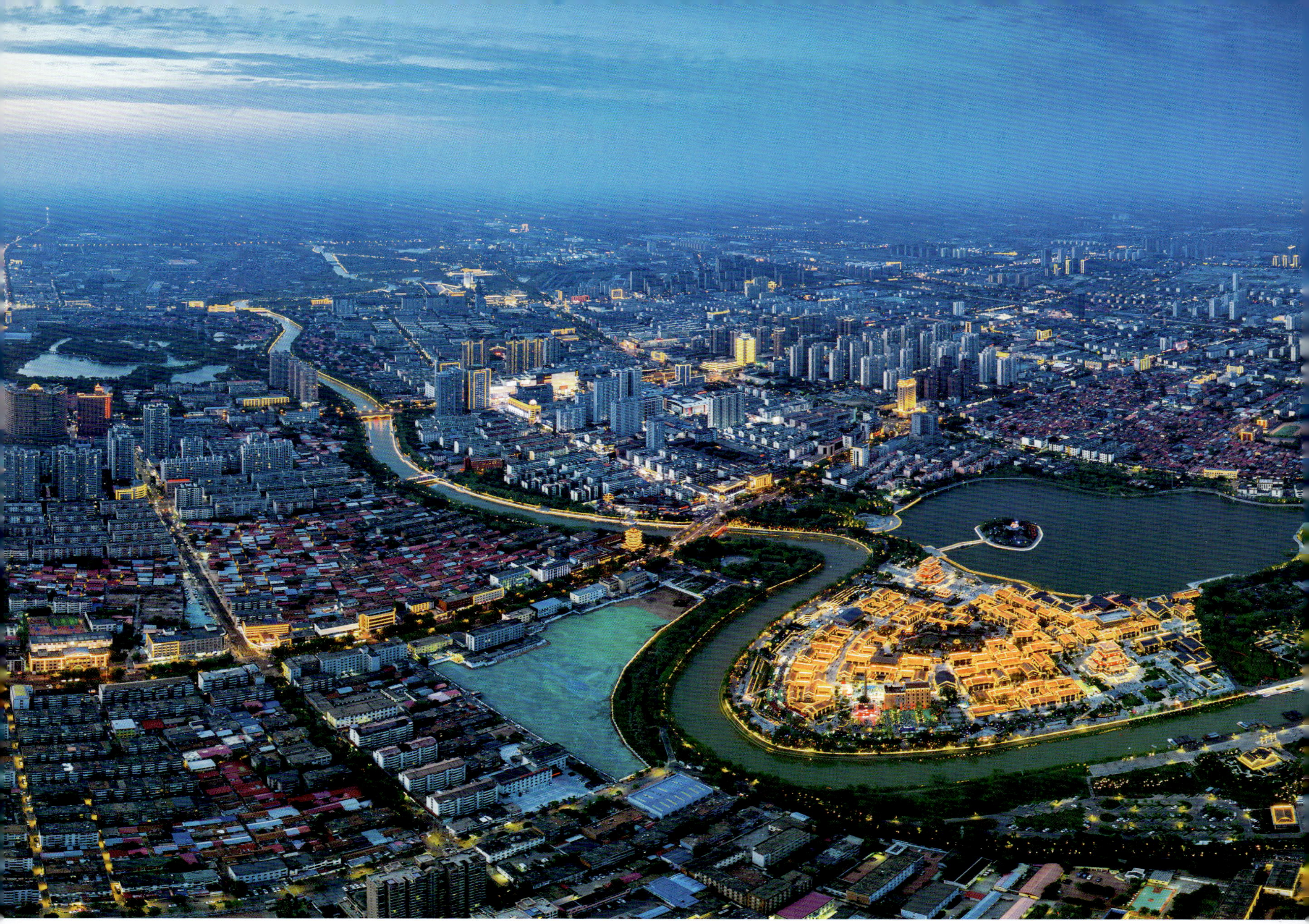

京杭大运河河北沧州段

沧州是京杭大运河流经里程最长的城市。在市该境内，大运河全长 216 千米，约占京杭大运河总长度的 1/7。

《璀璨狮城》，大运河沧州市区段，摄影：陈秀峰

The Cangzhou Section of the Beijing-Hangzhou Grand Canal, Hebei Province

Cangzhou is the city with the longest length of the Beijing-Hangzhou Grand Canal, with a total length of 216 kilometers, accounting for about one-seventh of the total length of the Grand Canal.

Brilliant Lion City, the Grand Canal in Cangzhou City
Photographed by Chen Xiufeng

潮白河河北段

潮白河为中国海河水系五大河之一，贯穿河北省、北京市、天津市三省市，全长 467 千米，流域面积约为 19354 平方千米，是北京市重要水源之一。
《廊坊潮白河》，潮白河廊坊市段，摄影：李辉

The Hebei Section of the Chaobai River
It's one of the five major rivers in the Haihe River system in China, running through Hebei Province, Beijing and Tianjin. The Chaobai River has a total length of 467 kilometers and a drainage area of 19,354 square kilometers, it is one of the important water sources for Beijing.
Langfang Chaobai River, Langfang City section of Chaobai River
Photography by Li Hui

滹沱河河北段

滹沱河是海河水系子牙河的上游支流之一，全长587千米，流域面积2.73万平方千米。该河流以泥沙多、善冲、善淤、善徙而闻名于北方。
《大美滹沱河》，滹沱河石家庄市段，摄影：封玉梅

The Hebei Section of the Hutuo River
It is one of the upstream tributaries of Ziya River in Haihe River system, with a total length of 587 kilometers and a drainage area of 27,300 square kilometers. The main stream is famous in the north for its high sediment load and its proneness to frequent scouring, siltation, and changes in the river course.
Magnificent Hutuo River, Shijiazhuang City section of the Hutuo River
Photographed by Feng Yumei

辽河源头

　　辽河源头位于平泉市七老图山脉的光头山附近。此处为辽河干流上游老哈河的发源地,附近有辽河源国家森林公园。
《辽河源头》,平泉市柳溪镇,摄影:杨树海

The source of Liaohe River, Hebei Province
Source of Liaohe River is located near the Guangtou Mountain in Qilaotu Mountains of Pingquan City, it is the birthplace of Laohahe River in the upper reaches of the Liaohe River. There is Liaoheyuan National Forest Park nearby.
The Source of the Liaohe River, Liuxi Town, Pingquan City
Photographed by Yang Shuhai

河北秦皇岛燕塞湖

燕塞湖因地处燕山要塞而得名。其地理位置在山海关城西 3.5 千米处，为万里长城东起的第一座人工湖。湖区面积约为 15 平方千米，是长寿山国家森林公园和秦皇岛柳江国家地质公园的重要组成部分。
《燕塞湖》，秦皇岛市山海关区，摄影：袁晓虹

Qinhuangdao Yansai Lake , Hebei Province

It's named after its location in the Yanshan Fortress. Its geographical location is 3.5 kilometers west of Shanhaiguan City. It's the first artificial lake from the east of the Great Wall. The area of this lake is about 15 square kilometers, and there is an important part of the Changshou Mountain National Forest Park and Qinhuangdao Liujiang National Geopark.
Yansai Lake, Shanhaiguan District, Qinhuangdao City
Photographed by Yuan Xiaohong

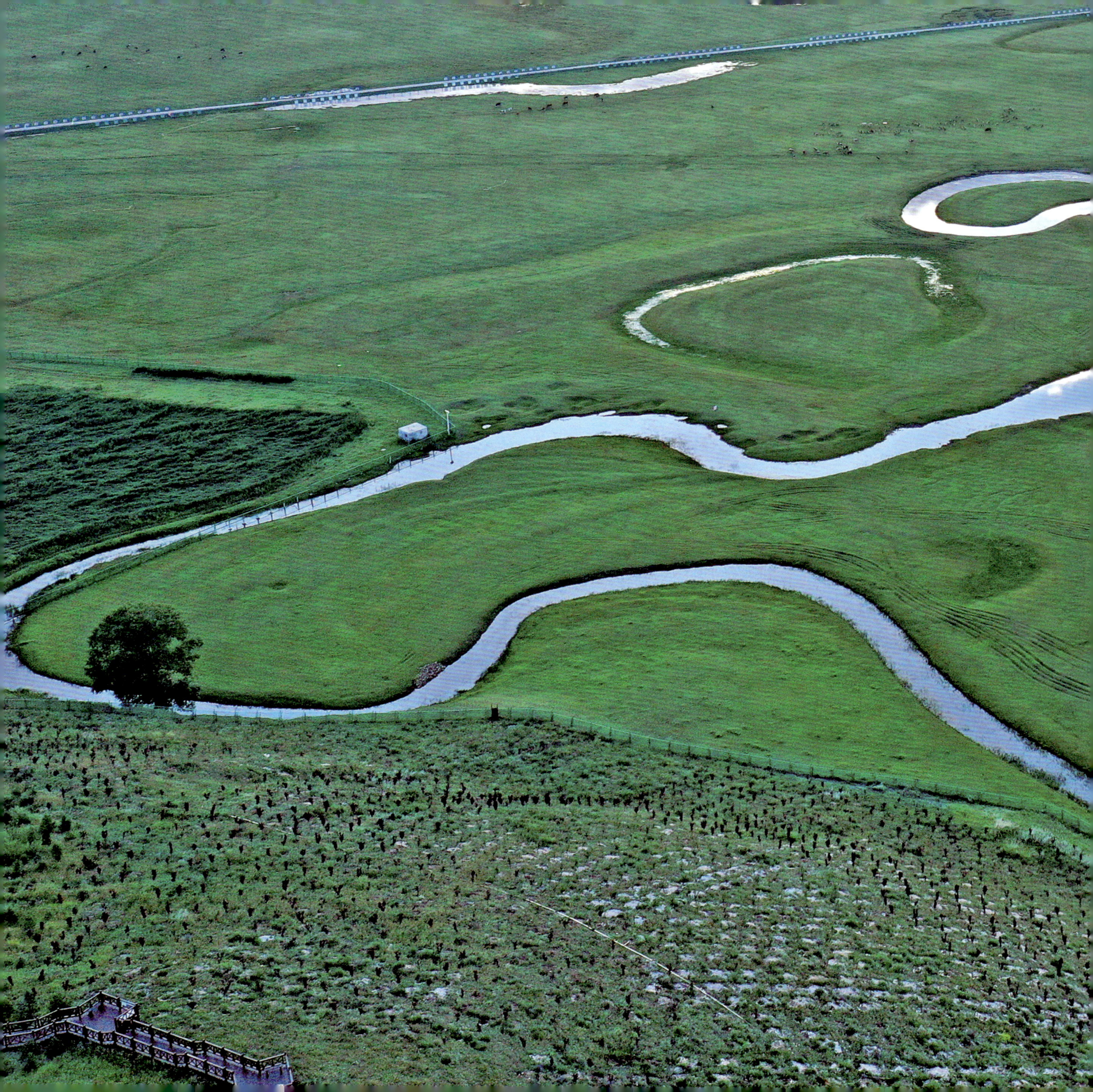

闪电河河北段

闪电河是滦河的源头，发源于沽源与丰宁满族自治县两县交界的古巴颜屯图固尔山。该河流蜿蜒曲折，全长 877 千米，最终注入渤海。
《闪电河神韵》，张家口市沽源县，摄影：厉文中

Hebei Section of the Lightning River
It is the source of the Luanhe River, originating from the Guba Yan Tun Tuguer Mountain at the border between Guyuan County and Fengning Manchu Autonomous County. The river meanders and twists, with a total length of 877 kilometers, and ultimately flowing into the Bohai Sea.
Lightning River Charm, Guyuan County, Zhangjiakou City
Photography by Li Wenzhong

海洋资源

河北地处环渤海海岸核心地带，面朝大海，向海而生。沿海地区毗邻京津，是华北、东北及西北广阔地区进入太平洋、通向世界最便捷的出海口之一。

河北省大陆海岸线长约487千米，管辖海域7200多平方千米，分别约占全国的3%和2%。河北省海洋区位条件独特，现有3个沿海市和11个沿海县（市、区）、7个经济开发区。

河北省海洋资源丰富，有30多种具有较高经济价值的海洋生物。丰富的海草床生态系统主要分布在曹妃甸龙岛西北侧浅水海域，是我国北方面积最大的鳗草海草床。

河北沿海岸滩生态系统主要有基岩质海岸生态系统、砂质海岸生态系统、淤泥质海岸生态系统三类。

河北省拥有渤海湾、唐山湾及秦皇岛湾3个海湾，33个河口。这里现存5个海岛，均为已开发利用海岛，海岛陆域面积约36平方千米。

河北省历史悠久，是华夏文明的发祥地之一。在其漫长的海岸线上分布着众多人文建筑景区。其中，山海关老龙头的海上长城、秦皇岛市秦皇求仙入海处等景点久负盛名；北戴河鸽子窝公园闻名天下，是我国著名的海上观日出最佳地之一。

河北省海洋经济资源类型多样，海岸线开发程度较高，主要有渔业、盐业、交通运输业、工业、旅游业、矿产与能源。近年来，海上光伏、海上风电有新的发展和突破。

河北省沿海地区有曹妃甸等优良港资源，有0.15万平方千米滩涂和盐碱地，目前已建成的港口有秦皇岛港、唐山港、黄骅港、曹妃甸港。

Marine Resources

Hebei is located in the core area of the Bohai Rim, facing the sea and survive by relying on the ocean. The coastal area is adjacent to Beijing and Tianjin, and is one of the most convenient outlets for the vast areas of North China, Northeast China and Northwest China to enter the Pacific Ocean and the world.

Hebei Province has a coastline of about 487 kilometers and a sea area of more than 7,200 square kilometers, accounting for about 3% and 2% of the country, respectively. Hebei Province has unique marine location conditions, with 3 coastal cities, 11 coastal counties (cities and districts) and 7 economic development zones.

Hebei Province is rich in marine resources, with more than 30 species of marine organisms with high economic value. The typical ecosystem of abundant seagrass beds is mainly distributed in the shallow waters on the northwest side of Caofeidian Long Island, which is the largest eelgrass seagrass bed in northern China.

There are three main types of coastal beach ecosystems along Hebei Province: bedrock coastal ecosystem, sandy coastal ecosystem, and silty coastal ecosystem.

Hebei Province has 3 bays and 33 estuaries, including Bohai Bay, Tangshan Bay and Qinhuangdao Bay. There are 5 existing islands, all of which have been developed and utilized, with a land area of about 36 square kilometers.

Hebei Province has a long history and is one of the birthplaces of Chinese civilization. There are many cultural and architectural scenic spots along the long coastline. Among them, the Great Wall on the sea of the Old Dragon's Head of Shanhaiguan and the estuary where Emperor Qin Shi Huang sought elixirs in Qinhuangdao City have long been famous. Beidaihe Pigeon Nest Park is famous all over the world and is one of the best places to watch the sunrise on the sea in China.

Hebei Province has a variety of marine economic resources and a high degree of coastline development, mainly including fishery, salt industry, transportation, industry, tourism, minerals and energy. In recent years, there have been new developments and breakthroughs in offshore photovoltaic and offshore wind power.

The coastal areas of Hebei Province have excellent port site resources such as Caofeidian, with 1,500 square thousand of tidal flats and saline-alkali land, and the ports that have been built are Qinhuangdao Port, Tangshan Port, Huanghua Port and Caofeidian Port.

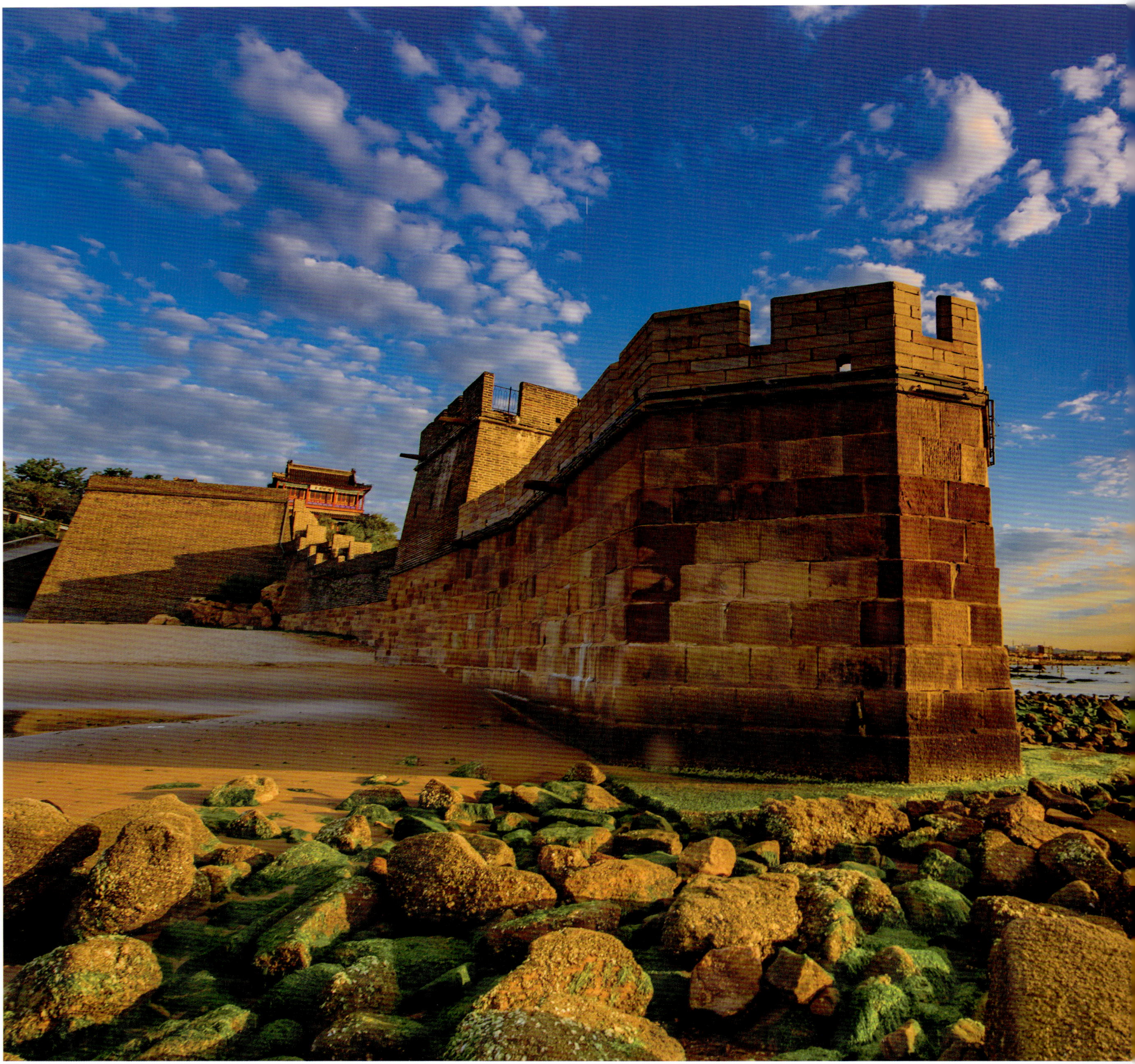

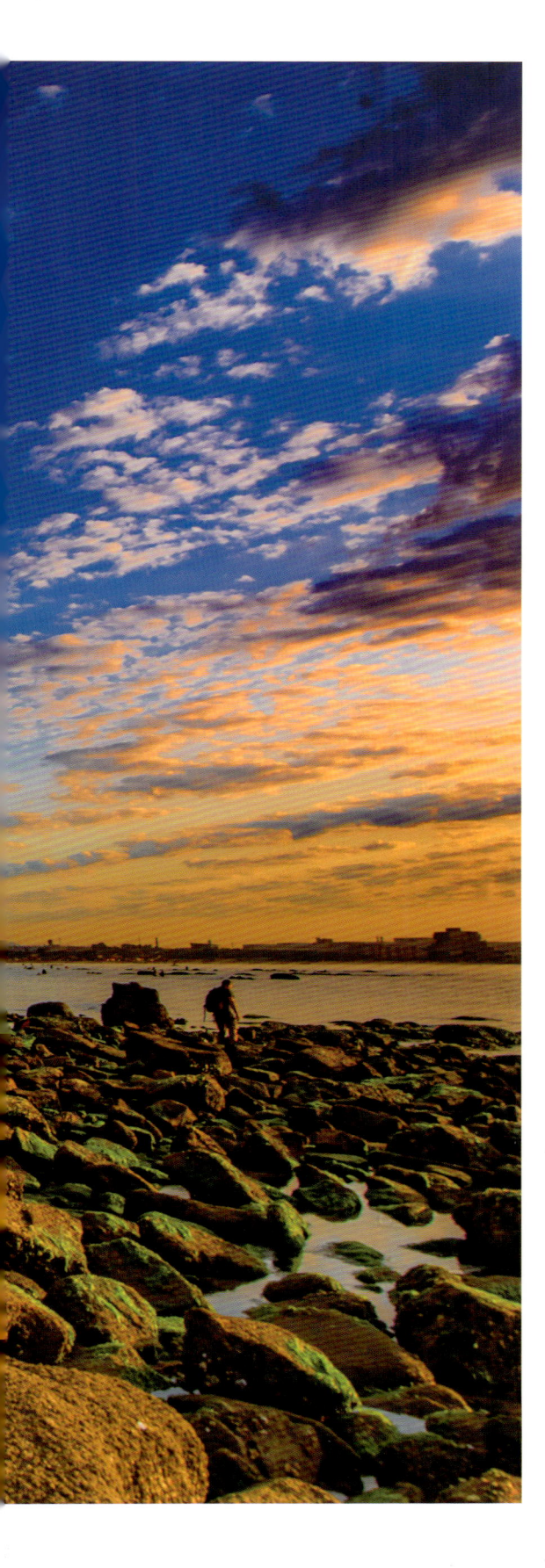

河北省海洋生态环境保护主要指标

Main indicators of marine ecological environment protection in Hebei Province

序号 Serial number	指标 Lndicatorw		指标类别 Lndicator category	"十三五"末现状 Current situation at the end of the 13th Five Year Plan period	2025年 In 2025
1	海洋环境质量 Marine	近岸海域优良（一、二类）水质比例（%） Proportion of excellent (Class I and II) water quality in nearshore waters (%)	约束性 Constrained nature	95.6（2018-2020年均值） 95.6 (average from 2018 to 2020)）	98 ninety-eight
2		旅游旺季北戴河主要海水浴场水质 During the peak tourist season, the water quality of the main beaches in Beidaihe is high	约束性 Constrained nature	一类 One type	一类 One type
3		入海河流国控断面达Ⅲ类水质比例（%） Proportion of Class III water quality in nationally controlled sections of rivers flowing into the sea (%)	预期性 Expectancy	46.2 forty-six point two	100（力争） 100 (Strive for)
4		省内国控入海河流总氮浓度（mg/L） Total nitrogen concentration of nationally controlled rivers entering the sea within the province (mg/L)	预期性 Expectancy	—	负增长（宣惠河下降5%） Negative growth (5% decline in
5	海洋生态修复 ocean ecology repair	大陆自然岸线保有率（%） Natural shoreline retention rate in mainland China (%)	约束性 Constrained nature	—	不减少 Not decreasing
6		岸线生态修复长度（千米） Length of shoreline ecological restoration (km)	预期性 Expectancy	—	≥30
7		退养还滩退围还海面积（公顷） Area of retiring, returning beaches, reclaiming fences, and returning seas (hectares)	预期性 Expectancy	—	≥4500
8		滨海湿地生态修复面积（公顷） Ecological restoration area of coastal wetlands (hectares)	约束性 Constrained nature	—	≥560
9		省级湿地公园（个） Provincial Wetland Park (s)	预期性 Expectancy	3 three	4 four
10		海草床养护面积（公顷） Seaweed bed maintenance area (hectares)	预期性 Expectancy	—	300 three hundred
11		文昌鱼平均栖息密度（个／平方米） Average habitat density of Wen Chang fish (individuals/square meter)	预期性 Expectancy	56 fifty-six	维持稳定 Maintain stability
12	亲海环境品质 Inhai environment quality	整治修复亲海岸滩长度（千米） Remediation and restoration of coastal beach length (km)	预期性 Expectancy	—	≥6
13		建设"美丽海湾"湾段数量（个） Number of "Beautiful Bay" bay sections constructed (units)	预期性 Expectancy	0	4 four

注：1. 到2025年秦皇岛、唐山、沧州近岸海域优良（一、二类）水质比例分别达到100%、99%、95%；
2. 预期性指标由重点工程项目汇总而来。

Note: 1 By 2025, the proportion of excellent (Class I and Class II) water quality in the coastal waters of Qinhuangdao, Tangshan, and Cangzhou will reach 100%, 99%, and 95% respectively;
2. Expected indicators are determined by weight Summary of engineering projects.

《万里长城老龙头》，万里长城山海关老龙头景区，摄影：田利君

The Old Dragon's Head of the Great Wall, The Great Wall of China Shanhaiguan Old Dragon Head Scenic Area
Photographed by Tian Lijun

河北秦皇岛

秦皇岛作为河北省三个沿海市之一,是国务院确定的中国著名滨海旅游、休闲、度假胜地,也是环渤海地区重要的港口城市。
《美丽北戴河》,秦皇岛市北戴河新区蔚蓝海岸,摄影:崔重辉

Qinhuangdao, Hebei Province
Qinhuangdao is one of the three coastal cities in Hebei province and an important port city in the Bohai Sea region. It is a renowned seaside resort approved by the State Council for tourism, recreation and holidays.
Beautiful Beidaihe, The Cote d'Azur, Beidaihe New District, Qinhuangdao City
Photographed by Cui Chonghui

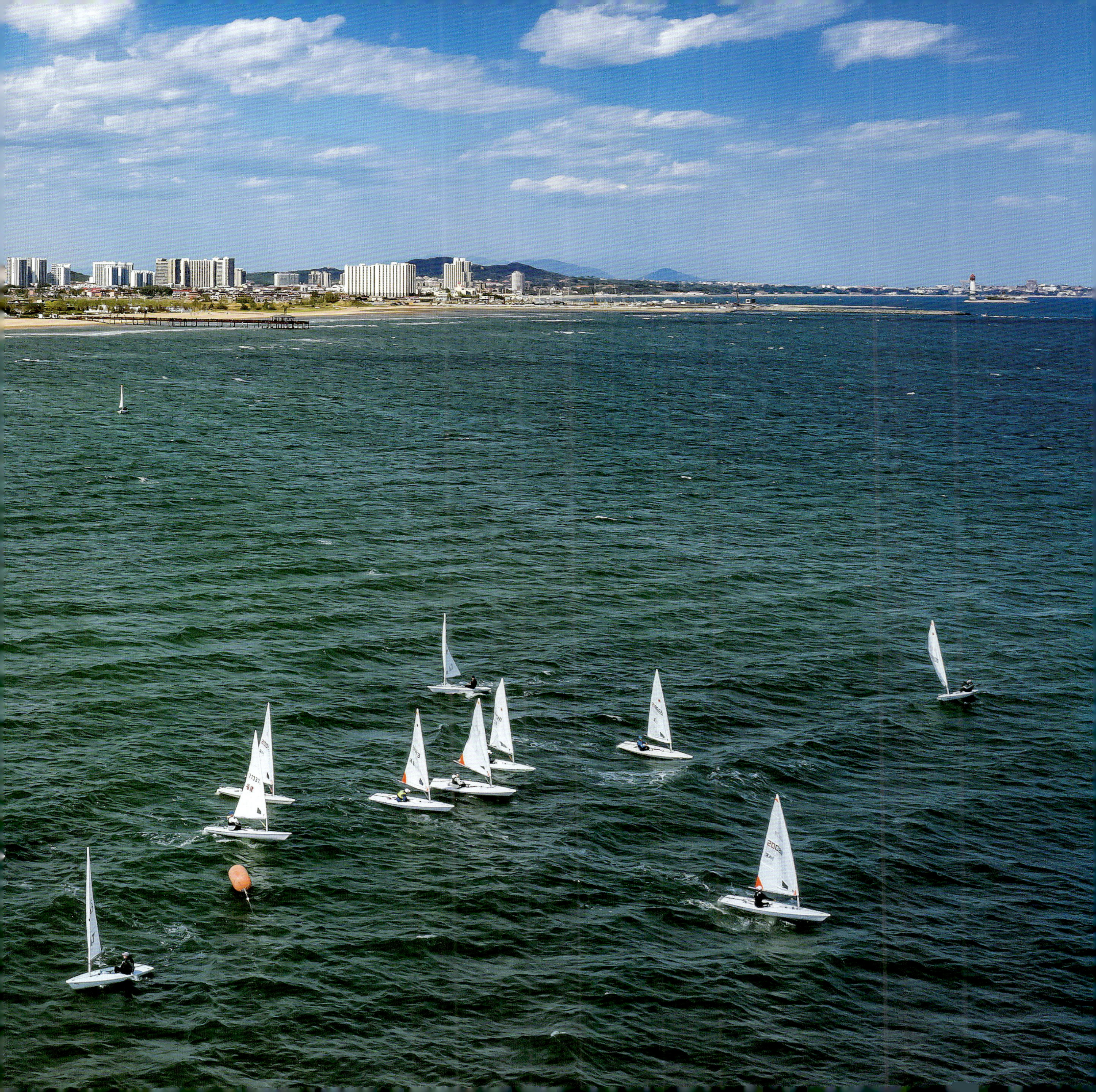

河北唐山

唐山作为河北省三个沿海市之一,是国务院确定的河北省中心城市之一,也是环渤海地区新型工业化基地和港口城市,中国(唐山)跨境电子商务综合试验区、中国(河北)自由贸易试验区组成部分。
《三贝明珠码头》,唐山市乐亭县,摄影:徐小跃

Tangshan, Hebei Province
As one of the three coastal cities in Hebei province, Tangshan is approved one of the central cities of Hebei province by the State Council. It is a new industrialisation base and port city in the Bohai Rim region, and also a part of China (Tangshan) Cross-Border E-Commerce Comprehensive Pilot Zone and China (Hebei) Pilot Free Trade Zone.
Sanbei Pearl Pier, Laoting County, Tangshan City
Photographed by Xu Xiaoyue

河北沧州

沧州作为河北省三个沿海市之一，在历史上曾是"海上丝绸之路"的北方起点。
《沧州海岸》，沧州市，供图：视觉中国

Cangzhou, Hebei Province
As one of the three coastal cities in Hebei province, Cangzhou was historically the northern starting point of the Maritime Silk Road.
Coastal Cangzhou, Cangzhou City
Photo provided by Visual China

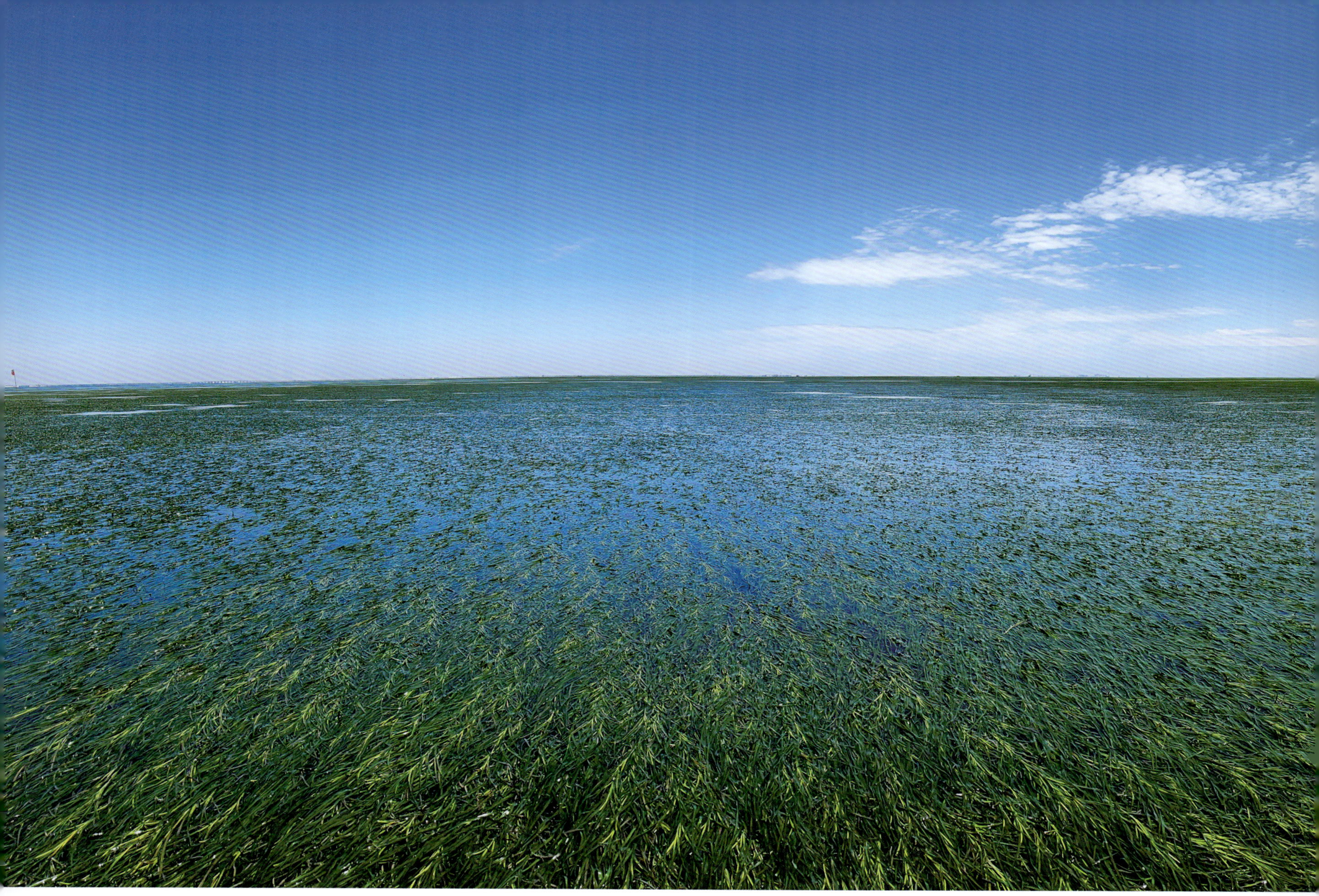

河北海草床生态修复工程

河北省海草床生态系统主要分布在曹妃甸龙岛西北侧浅水海域。此处的海草种类为鳗草，是我国北方面积最大的鳗草海草床。

《海草床生态修复工程》，唐山市曹妃甸区龙岛，摄影：左立明

Seagrass Bed Ecological Restoration Project in Hebei province

The typical ecosystem of seagrass bed in Hebei province is mainly distributed in the shallow sea area of the northwest of Caofeidian Long Island. It has the largest eelgrass seagrass bed in north China.

Seagrass Bed Ecological Restoration Project, Long Island in Caofeidian District, Tangshan City
Photographed by Zuo Liming

河北唐山国际旅游岛——菩提岛

　　菩提岛为唐山国际旅游岛组成部分。这里紧邻秦皇岛市北戴河新区。菩提岛上有各种植物260多种，自然植被覆盖率达98%以上，有"孤悬于海上的天然动植物园"之美誉。此外，这里还有潮音寺和朝阳庵遗址等佛教古迹。

《菩提岛》，唐山市乐亭县，摄影：关欣

Tangshan Bay International Tourism Island, Hebei Province - Bodhi Island

Bodhi Island is one of the components of Tangshan Bay International Tourism Island. It near Beidaihe new district, Qinhuangdao city. With more than 260 species of plants, and coverage rate over 98% of the natural vegetation, it enjoys the reputation of being a "natural zoological and botanical garden hanging over the sea". In addition, there are Buddhist monuments such as Chaoyin Temple and the ruins of Chaoyangan.

Bodhi Island, Laoting County, Tangshan City
Photographed by Guan Xin

河北唐山国际旅游岛——月岛

月岛为唐山国际旅游岛的组成部分，位于唐山市乐亭县西南侧的渤海湾中，是一处沙质海岛。这座小岛因其形状像一弯月亮而得名。月岛上沙丘松软，滩缓潮平，景色优美。岛上一幢幢荷兰风情的小木屋和彩色的别墅更为小岛增添了几分迷人风采，使其成为北京周边的"小马尔代夫"。
《月岛》，唐山市乐亭县，摄影：杨建萍

Tangshan Bay International Tourism Island, Hebei Province, Moon Island
Moon Island is one of the components of Tangshan International Tourism Island, located in the Bohai Bay on the southwest side of Laoting County, Tangshan City, is a sandy island. The island gets its name from the fact that it is shaped like a curved moon. The sand dunes on Moon Island are soft, the beach is gentle and the tide is flat, and the scenery is beautiful. The island's Dutch-style chalets and colourful villas add to the charm of the island, making it the "Little Maldives" around Beijing.
Moon Island, Laoting County, Tangshan City
Photographed by Yang Jianping

河北唐山国际旅游岛——祥云岛

　　祥云岛为唐山国际旅游岛组成部分之一，位于唐山市海港开发区西南 5 千米的渤海湾黄金海岸处。从高空俯视该岛，形状好似保龄球。祥云岛是我国最大的由河流和海汐冲积而成的细沙岛屿，也是附近沿海旅游风景线上有待开发的最大亮点。
《祥云岛》，唐山市乐亭县，摄影：张欢

Tangshan Bay International Tourism Island, Hebei Province - Xiangyun Island
Xiangyun Island is one of the components of Tangshan International Tourism Island, located on the Gold Coast of Bohai Bay, 5 kilometers southwest of Tangshan Harbor Development Zone. Looking down on the island from above, it looks like a bowling ball. Xiangyun Island is the largest fine sand island formed by rivers and tidal alluvials in China, which is also the biggest highlight to be developed on the nearby coastal tourist landscape.
Xiangyun Island, Laoting County, Tangshan City
Photographed by Zhang Huan

■ 龙岛

　　龙岛横卧在曹妃甸海域老龙沟东侧，鸟瞰龙岛，如同阿拉伯数字"7"。这里是一座古滦河入海冲积而成的岛屿。该岛为海洋中未完全开发的原始孤岛，荒野韵味十足，是渤海湾中一块珍贵的处女地和未经雕琢的天然玉带。
《龙岛》，唐山市曹妃甸区，摄影：朱米福

Long Island
Long Island lies on the east side of the Laolonggou in the Caofeidian Sea area, and when viewed from the air, its shape resembles the Arabic figure "7". Created by the alluvial flow of the ancient Luanhe River into the sea, it is a primitive island in the sea, full of wild charm, and a precious virgin land and uncarved natural jade belt in the Bohai Bay.
Long Island, Caofeidian District, Tangshan City
Photographed by Zhu Mifu

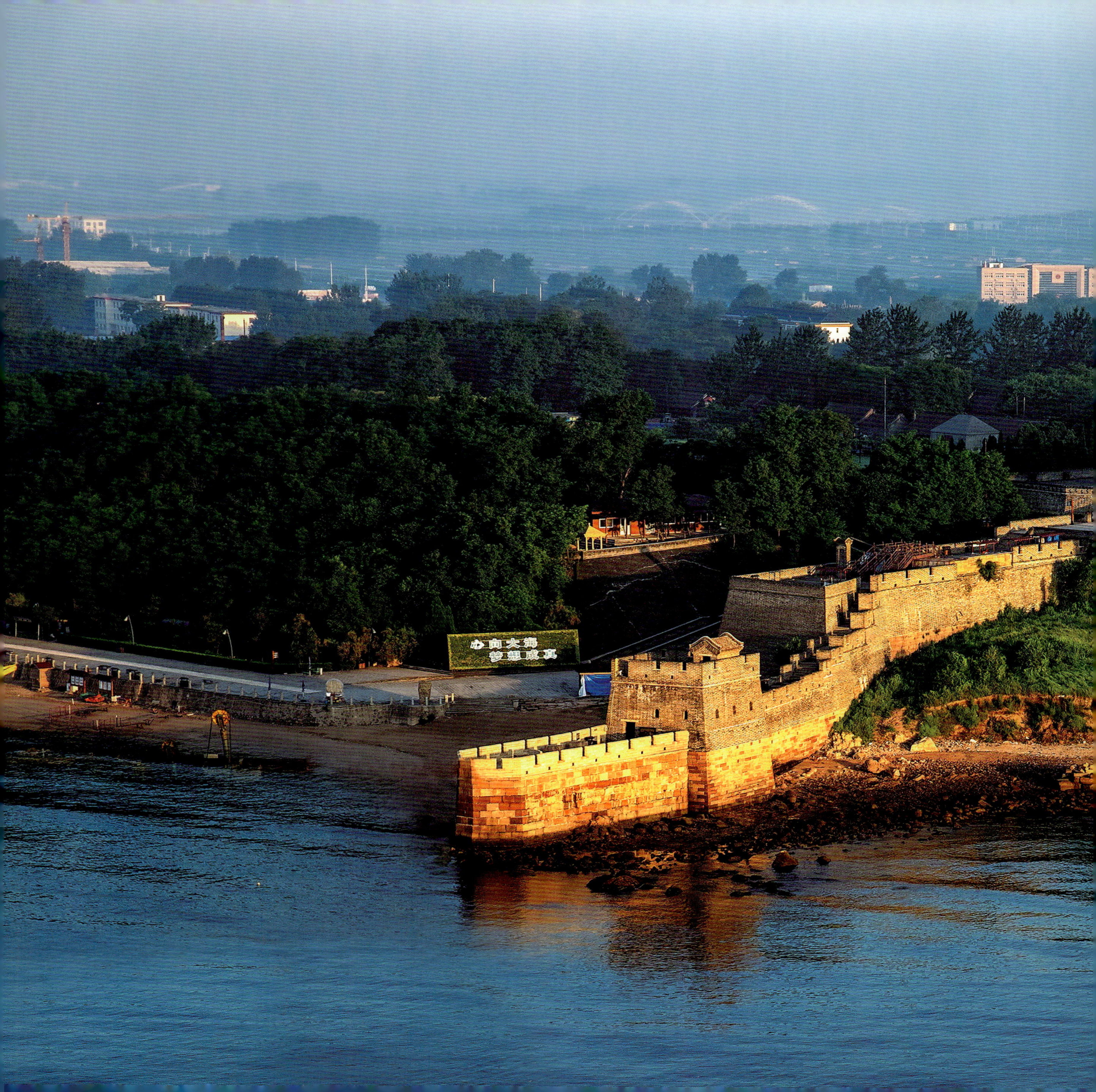
心向大海
梦想成真

河北秦皇岛老龙头

老龙头位于秦皇岛市山海关区城南5千米处，是明长城的东端入海处。这里最为著名的建筑当属有"长城连海水连天，人上飞楼百尺巅"之称的澄海楼。
《万里长城老龙头》，秦皇岛市老龙头风景区，摄影：朗晓光

Old Dragon's Head in Qinhuangdao, Hebei Province

Old Dragon's Head, located in Shanhaiguan district of Qinhuangdao city, 5 kilometers south of the city. It is the eastern entrance to the sea of the Ming Great Wall. Its most famous component is Chenghai Tower, which enjoys the reputation that "the Great Wall stretches into the sea and the sea merges with the sky; those who climb the tower feel as if they are hundreds of meters above the ground".
The Old Dragon's Head of Great Wall, Qinhuangdao City Old Dragon Head Scenic Area
Photographed by Lang Xiaoguang

河北秦皇岛秦皇求仙入海处

秦皇求仙入海处位于秦皇岛市海港区东南部,因秦始皇东巡驻跸于此而得名。作为以秦皇求仙入海和战国文化为主题的文化旅游区,这里的主要景观有阙门区、战国风情区、求仙区、沿海区四部分。

《秦皇求仙入海处》,秦皇岛市海港区,摄影:田利君

The estuary where Emperor Qin Shi Huang sought elixirs, Qinhuangdao, Hebei Province
The estuary where Emperor Qin Shi Huang sought elixirs is located in the southeast Haigang District, Qinhuangdao City, which is named for China's first emperor Qin Shihuang who once rested here on his eastern tour. As a cultural tourism area with the theme of the Emperor Qin seeking immortals for the way to immortality into the sea and the Warring States culture, its main landscape consists of four parts: the Quemen area, the Warring States scenery area, the fairy area and the coastal area.

The Estuary Where Emperor Qin Shi Huang Sought Elixirs, Haigang District, Qinhuangdao City
Photographed by Tian Lijun

河北北戴河鸽子窝公园

鸽子窝公园位于北戴河海滨的东北角，是北戴河风景名胜区四大景区之一。20世纪90年代末，这里定为国家级鸟类自然保护区。在鸽子窝观日出时还常常可见到"浴日"的奇景。1954年夏，毛主席曾在此写下《浪淘沙·北戴河》。

《美丽的鸽子窝公园》，秦皇岛市北戴河海滨，摄影：崔重辉

Beidaihe Pigeon Nest Park, Hebei Province

Pigeon Nest Park is located in the northeast corner of Beidaihe coast. This is one of the four best tourist attractions of Beidaihe Scenic Area. In the late 1990s, it was designated as a national level bird nature reserve. Watching the sunrise over the pigeon nest is often a great way to see the miracle of the "bathing sun". In the summer of 1954, Chairman Mao wrote "Waves of Sand, Beidaihe" here

Beautiful Pigeon Nest Park, Beidaihe Seaside in Qinhuangdao City
Photographed by Cui Chonghui

河北秦皇岛港

秦皇岛港位于渤海辽东湾西侧，滨海平原的东北侧，靠近万里长城的东端，地处山海关要冲。这里港阔水深，风平浪小，一年四季不冻不淤，是中国北方的一座天然港口。
《秦皇岛港》，黄骅市境内港口，摄影：崔重辉

Port of Qinhuangdao, Hebei Province
Lying on the west side of Liaodong Bay of the Bohai Sea, the northeast side of the coastal plain, near the eastern end of the Great Wall, Port of Qinhuangdao is at the key point of the Shanhai Pass. The harbor here is wide and deep, with shallow wind and shallow waves, no ice and no siltation all year round, it is a natural harbour in northern China.
Port of Qinhuangdao, Port in Huanghua City
Photographed by Cui Chonghui

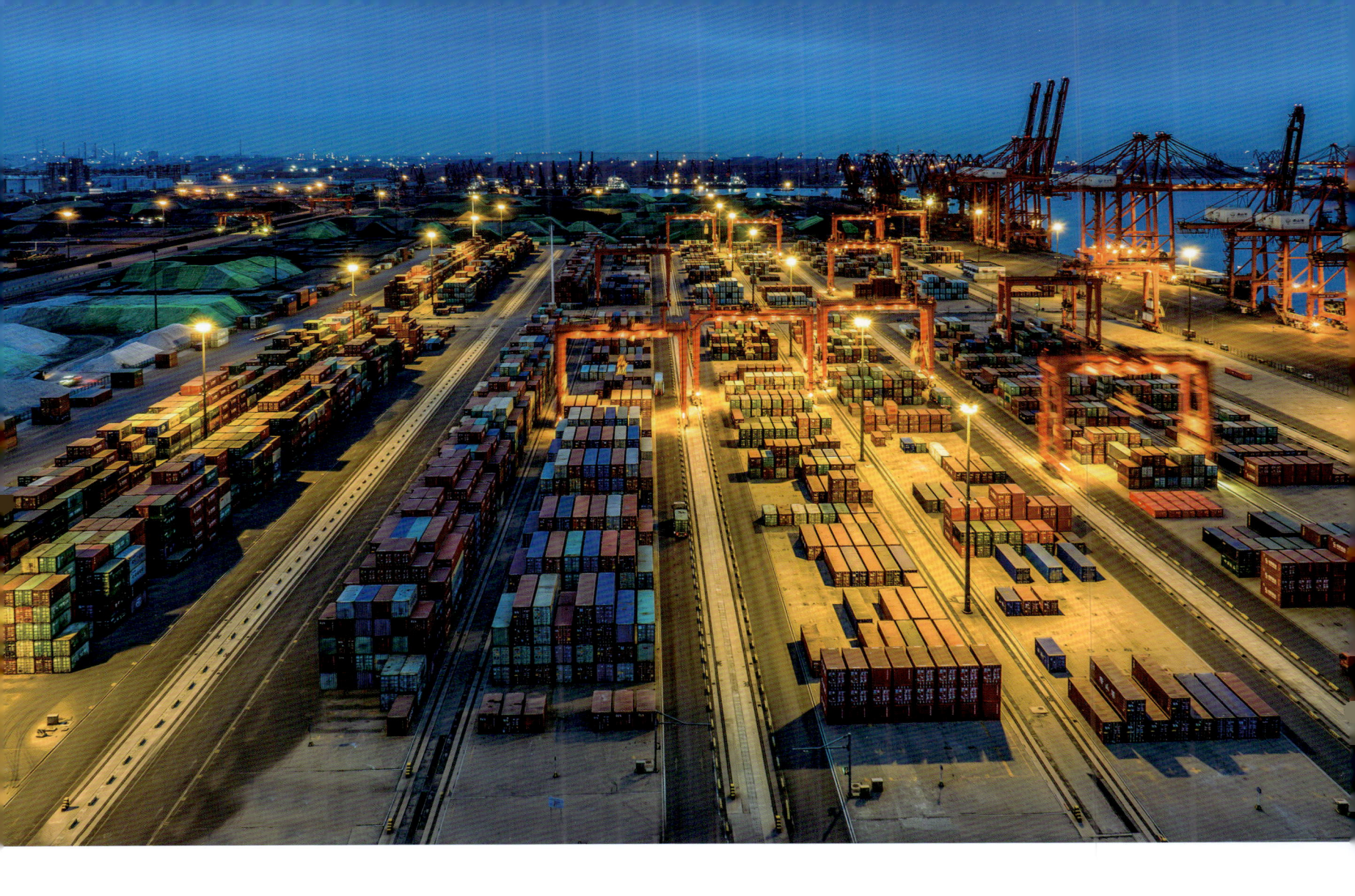

河北唐山港

唐山港位于唐山市东南沿海，为曹妃甸循环经济示范区、唐山海港经济开发区的重要基础设施及战略资源。2022 年，唐山港全港完成货物吞吐量 76887 万吨，全年货物吞吐量居世界港口第二位。

《唐山港》，唐山市东南沿海，摄影：杨建萍

Port of Tangshan, Hebei Province
The Port of Tangshan is located in the southeast coast of Tangshan City, it is an important infrastructure and strategic resource for the development and construction of Caofeidian Circular Economy Demonstration Zone and Tangshan Seaport Economic Development Zone. In 2022, it completed 768.87 million tons of cargo throughput, ranking second in the world in terms of annual cargo throughput.

The Port of Tangshan, Southeast coast of Tangshan City
Photographed by Yang Jianping

河北曹妃甸港

曹妃甸港位于滦河三角洲西侧区域。这里具有水深、无淤积（不需开挖）、不冻、近陆地、交通便捷、滩涂宽阔、造地成本低七大优势。"面向大海有深槽，背靠陆地有浅滩，地下储有大油田"是曹妃甸最明显的自然地理特征。
《曹妃甸港》，滦河三角洲，摄影：魏伟

Port of Caofeidian, Hebei Province
Caofeidian Port is located in the western area of the Luanhe Delta. There are seven advantages: deep water, no silting (no excavation), no ice, close to the land, convenient transportation, wide beach, and low land construction cost. "Deep trough facing the sea, shoal backing the land, and large underground oil reservoir" are the most obvious natural geographical features of Caofeidian.
Port of Caofeidian, Luanhe River Delta
Photographed by Wei Wei

河北黄骅港

黄骅港位于沧州黄骅市的渤海之滨，现已建成20万吨级航道和万吨级以上泊位25个，吞吐量连续3年突破亿吨，成为亚欧大陆桥的新通道、桥头堡。
《黄骅港》，黄骅市，摄影：刘国昌

Port of Huanghua, Hebei Province
Port of Huanghua is located on the shore of the Bohai Sea in Huanghua, Cangzhou city. It has built a 200,000-ton waterway and 25 berths above 10,000-ton, with a throughput exceeding 100 million tons for three consecutive years, becoming a new channel and bridgehead of the Eurasian Land Bridge.
Port of Huanghua, Huanghua City
Photographed by Liu Guochang

野生动物资源

河北省总面积约为18.88万平方千米，是全国唯一兼有高原、山地、丘陵、平原、湖泊和海滨的省份，复杂的地形地貌及丰富的自然资源为众多的野生动物栖息繁衍提供了良好的生存环境。

近年来，河北省委、省政府高度重视陆生野生动物保护工作，修订了《河北省陆生野生动物保护条例》，颁布了《河北省湿地保护条例》，划定了野生动物禁猎区，规定了野生动物禁猎期等。自此，野生动物栖息地得到了有效的保护，陆生野生动物物种数量和种群数量得以恢复。

根据陆生野生动物资源调查和有关资料，河北省现有陆生野生脊椎动物605种，包括哺乳类87种，鸟类486种，两栖类8种，爬行类24种。其中，国家重点保护陆生野生动物142种，包括国家一级保护陆生野生动物华北豹、褐马鸡、黑鹳、东方白鹳、大鸨等43种，国家二级保护陆生野生动物豹猫、大天鹅、小天鹅、鸳鸯等99种，省级重点保护陆生野生动物狍、灰雁、豆雁等164种。

国家"十四五"期间抢救性保护的极度濒危物种中，分布于河北省的有：豹、梅花鹿、麋鹿、丹顶鹤、猎隼、黑脸琵鹭、大鸨、中华凤头燕鸥、中华秋沙鸭。其中、麋鹿、梅花鹿没有野外种群。

Wildlife Resources

With a total area of about 188,800 square kilometers, Hebei Province is the only province in the country that has plateaus, mountains, hills, plains, lakes and seashores. The complex topography and abundant natural resources provide a good living environment for many wild animals to inhabit and reproduce.

In recent years, the Hebei Provincial Party Committee and the Provincial Government have attached great importance to the protection of terrestrial wild animals, revised the *Regulations on the Protection of Terrestrial Wild Animals in Hebei Province*, promulgated *the Regulations on the Protection of Wetlands in Hebei Province*, demarcated wild animal hunting ban areas, and stipulated the hunting ban period for wild animals. Since then, wildlife habitats have been effectively improved, and the number of terrestrial wildlife species and populations have been protected and restored.

According to the survey of terrestrial wildlife resources and relevant data, there are 605 species of terrestrial wild vertebrates in Hebei Province, including 87 species of mammals, 486 species of birds, 8 species of amphibians and 24 species of reptiles. Among them, there are 142 species of terrestrial wild animals under national key protection, including 43 species of terrestrial wild animals under national first-class protection, such as the North China leopard, brown pheasant, black stork, Oriental white crane and great bustard, 99 species of terrestrial wild animals under national second-class protection, such as leopard cat, whooper swan, whistling swan and mandarin duck, and 164 species of terrestrial wild animals under provincial key protection, such as roe deer, gray goose and bean goose.

During the 14th Five-Year Plan period, the critically endangered species under rescue protection in Hebei Province include leopards, sika deer, elk, red-crowned cranes, falcons, black-faced spoonbills, great bustards, Chinese crested terns and Chinese mergansers. Among them, elk and sika deer have no wild populations.

梅花鹿（左上图）

分布范围：亚洲东部的特产种类，主要分布在俄罗斯东部、日本和中国。河北省围场、兴隆有分布，现野外种群少见。
国家一级保护陆生野生动物，摄影：王德庆

Sika Deer (*Cervus nippon*) Top left
Distribution: A specialist species of eastern Asia, it found mainly in eastern Russia, Japan and China. There are distributions in Weichang and Xinglong in Hebei province, but now the wild populations are rare.
National first-class protected terrestrial wildlife
Photographed by Wang Deqing

蒙原羚（左中图）

分布范围：国内分布在黑龙江、吉林、辽宁、内蒙古、山西、陕西、甘肃、宁夏等地。河北省见于北部坝上高原区。
国家一级保护陆生野生动物，摄影：赵俊

Mongolian Gazelle (*Procapra gutturosa*) Middle left
Distribution: In China, it is distributed in Heilongjiang, Jilin, Liaoning, Inner Mongolia, Shanxi, Shaanxi, Gansu, Ningxia and other provinces. In Hebei Province, found in the northern Bashang plateau area.
National first-class protected terrestrial wildlife
Photographed by Zhao Jun

黑嘴松鸡（下图）

分布范围：见于中国东北地区，可游荡至河北北部。
国家一级保护陆生野生动物，摄影：刘庆顺

Black-billed Capercaillie (*Tetrao urogalloides*) Figure below
Distribution: It is found in northeast China, wandering as far north as northern Hebei.
National first-class protected terrestrial wildlife
Photographed by Liu Qingshun

■ **中华秋沙鸭**（右上图）

分布范围：西伯利亚以及中国的黑龙江、吉林、河北等地。
国家一级保护陆生野生动物，摄影：王秀荣

Chinese Merganser (*Mergus squamatus*) Top right
Distribution: In Siberia and China's Heilongjiang, Jilin, Hebei and other places.
National first-class protected terrestrial wildlife
Photographed by Wang Xiurong

■ **中华凤头燕鸥**（右中图）

分布范围：在国外分布于泰国湾、加里曼丹岛北部沿海、菲律宾中部与北部沿海和岛屿以及马来半岛海岸等地。我国河北、天津、山东、上海、浙江、福建、台湾、广东、海南均有分布。
国家一级保护陆生野生动物，摄影：张玉成

Chinese Crested Tern (*Thalasseus bernsteini*) Middle right
Distribution: It is distributed overseas in the Gulf of Thailand, along the northern coast of Kalimantan Island, along the central and northern coasts and islands of the Philippines, and along the coast of the Malay Peninsula. In China, found in Hebei, Tianjin, Shandong, Shanghai, Zhejiang, Fujian, Taiwan, Guangdong and Hainan.
National first-class protected terrestrial wildlife
Photographed by Zhang Yucheng

■ **黑琴鸡**（下图）

分布范围：国内分布于东北、内蒙古、新疆等地区；河北省内见于坝上等地。
国家一级保护陆生野生动物，摄影：王振山

Black Grouse (*Lyrurus tetrix*) Figure below
Distribution: In China, it is distributed in Northeast, Inner Mongolia, Xinjiang and other areas. In Hebei Province, can be seen in Bashang and other areas.
National first-class protected terrestrial wildlife
Photographed by Wang Zhenshan

白冠长尾雉

分布范围:为我国特有鸟种,分布于河南、河北、陕西、山西、湖北、湖南、贵州、安徽等省。
国家一级保护陆生野生动物,摄影:胡琳

Reeves's Pheasant (*Syrmaticus reevesii*)

Distribution: A unique bird species in China, It is distributed in Henan, Hebei, Shaanxi, Shanxi, Hubei, Hunan, Guizhou, Anhui and other provinces.
National first-class protected terrestrial wildlife
Photographed by Hu Lin

大鸨

分布范围：广泛分布于欧亚大陆——从欧洲的伊比利亚半岛向东到土耳其、蒙古国、俄罗斯、中国以及朝鲜半岛等地。河北省坝上地区有繁殖，平山、白洋淀、宁晋及沿海地区为冬候鸟。
国家一级保护陆生野生动物，摄影：甘跃华

Great Bustard (*Otis tarda*)

Distribution: It is widely distributed in Eurasia, from the Iberian Peninsula in Europe eastward to Turkey, Mongolia, Russia, China and the Korean Peninsula and other areas in Asia. Breeds in Bashang area, as a wintering migratory bird in Pingshan, Baiyangdian, Ningjin and coastal areas of Hebei Province.
National first-class protected terrestrial wildlife
Photographed by Gan Yuehua

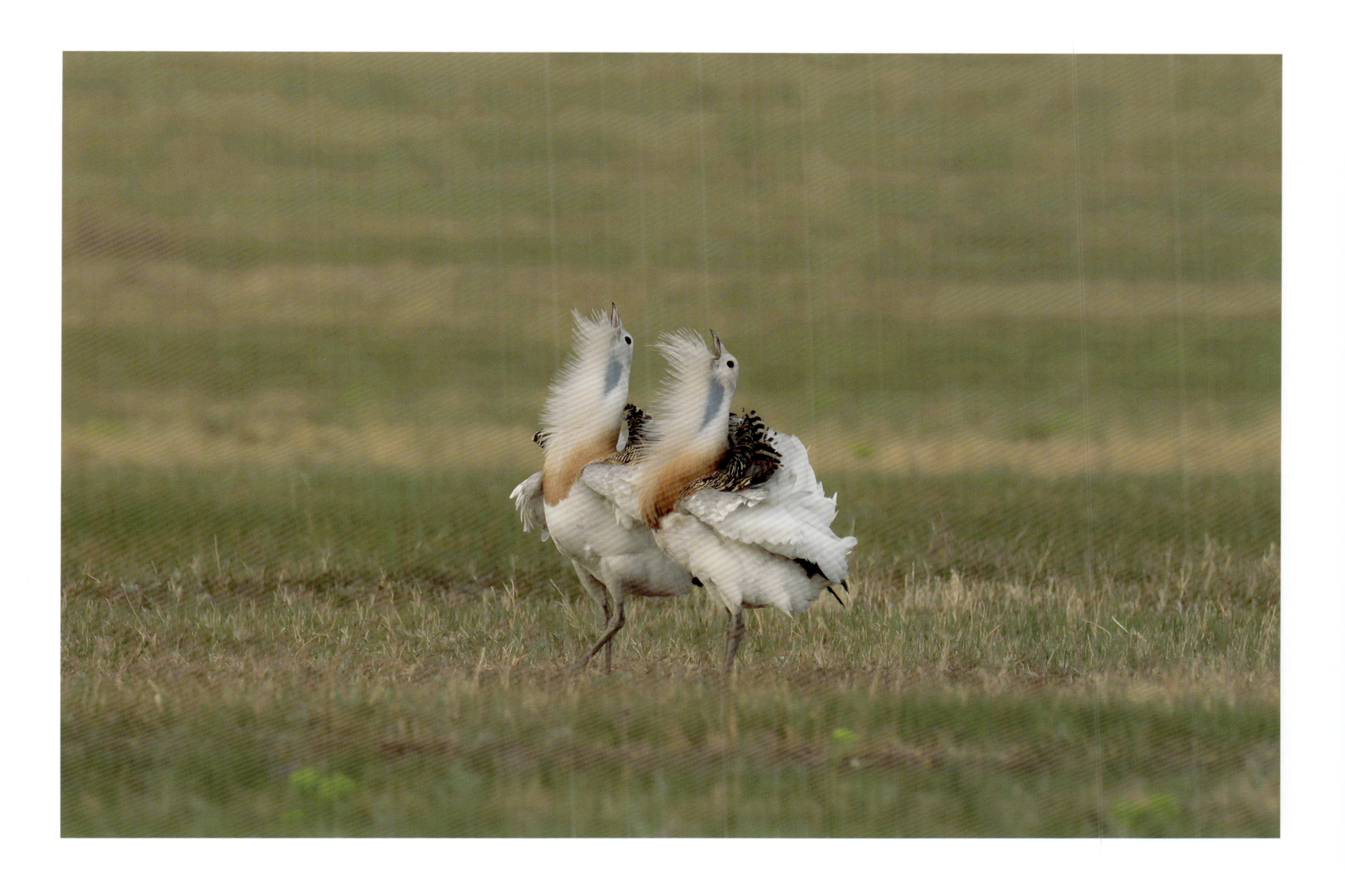

豺

分布范围：我国除台湾与海南以外，各地均有分布。河北省主要见于山区。
国家一级保护陆生野生动物，摄影：吴宗凯

Dhole (*Cuon alpinus*)
Distribution: It is distributed throughout China except Taiwan and Hainan. In Hebei Province, Mainly found in the mountainous areas.
National first-class protected terrestrial wildlife
Photographed by Wu Zongkai

■ 金钱豹

分布范围：国内各地均有分布。河北省见于山区森林中。
国家一级保护陆生野生动物，摄影：更尕索南

Leopard (*Panthera pardus*)
Distribution: It is distributed all over China. In Hebei Province, found in mountainous forests.
National first-class protected terrestrial wildlife
Photographed by Gengsuonan

褐马鸡（上图）

分布范围：国内仅见于山西芦芽山、管涔山国家森林公园，河北西北部，陕西黄龙山和北京灵山。河北省分布范围为小五台山及周边山区。在河北省为留鸟。
国家一级保护陆生野生动物，摄影：田峰

Brown Eared-pheasant (*Crossoptilon mantchuricum*) Above
Distribution: In China, it is only found in Luya Mountain and Guancen Mountain National Forest Park in Shanxi, northwestern Hebei, Huanglong Mountain in Shaanxi and Lingshan Mountain in Beijing. In Hebei Province, found in Xiaowutai Mountains and the surrounding mountains. As a resident bird in Hebei Province.
National first-class protected terrestrial wildlife, Photographed by Tian Feng

波斑鸨（左下图）

分布范围：分布于内蒙古西部、中部和新疆北部，在河北省有记录。
国家一级保护陆生野生动物，摄影：王秀荣

Macqueen's Bustard (*Chlamydotis macqueenii*) Bottom left
Distribution: It is distributed in western and central Inner Mongolia and northern Xinjiang. There are also distribution records in Hebei Province.
National first-class protected terrestrial wildlife
Photographed by Wang Xiurong

卷羽鹈鹕（右下图）

分布范围：繁殖于东欧至中亚，越冬于非洲北部、希腊、土耳其、印度及中国东南沿海等地。我国多地有分布，河北省主要分布于秦皇岛、唐山和沧州的沿海地区（各地记录的斑嘴鹈鹕可能多数为卷羽鹈鹕）。
国家一级保护陆生野生动物，摄影：王秀荣

Dalmatian Pelican (*Pelecanus crispus*) Bottom right
Distribution: It usually breeds ground ranges from Eastern Europe to Central Asia, and overwinters in northern Africa, Greece, Turkey, India and the southeast coast of China. In Hebei Province, mainly distributed in the coastal areas of Qinhuangdao, Tangshan and Cangzhou (most of the Spot-billed Pelican recorded in various places may be Dalmatian Pelican).
National first-class protected terrestrial wildlife
Photographed by Wang Xiurong

遗鸥（上图）

分布范围：繁殖地集中在蒙古国、哈萨克斯坦、俄罗斯和中国。河北省分布于张家口坝上湿地等地。

国家一级保护陆生野生动物，摄影：张玉成

Relict Gull (*Ichthyaetus relictus*) Above

Distribution: It usually breeds in Mongolia, Kazakhstan, Russia and China. In Hebei Province, distributed in Bashang wetland of Zhangjiakou city and other places.

National first-class protected terrestrial wildlife

Photographed by Zhang Yucheng

斑嘴鹈鹕（左下图）

分布范围：分布于中国华东及华南沿海——从江苏至广西、云南南部、海南岛。

国家一级保护陆生野生动物，摄影：王秀荣

Spot-billed Pelican (*Pelecanus philippensis*) Bottom left

Distribution: It is distributed in the coastal areas of East China and South China, from Jiangsu to Guangxi, southern Yunnan, Hainan Island.

National first-class protected terrestrial wildlife

Photographed by Wang Xiurong

彩鹬（右下图）

分布范围：分布于我国重庆、四川、云南、浙江、福建、台湾等地，在河北省偶见。

国家一级保护陆生野生动物，摄影：王秀荣

Greater Painted-snipe (*Rostratula benghalensis*) Bottom right

Distribution: It is distributed in Chongqing, Sichuan, Yunnan, Zhejiang, Fujian, Taiwan and other places, rare in Hebei Province.

National first-class protected terrestrial wildlife

Photographed by Wang Xiurong

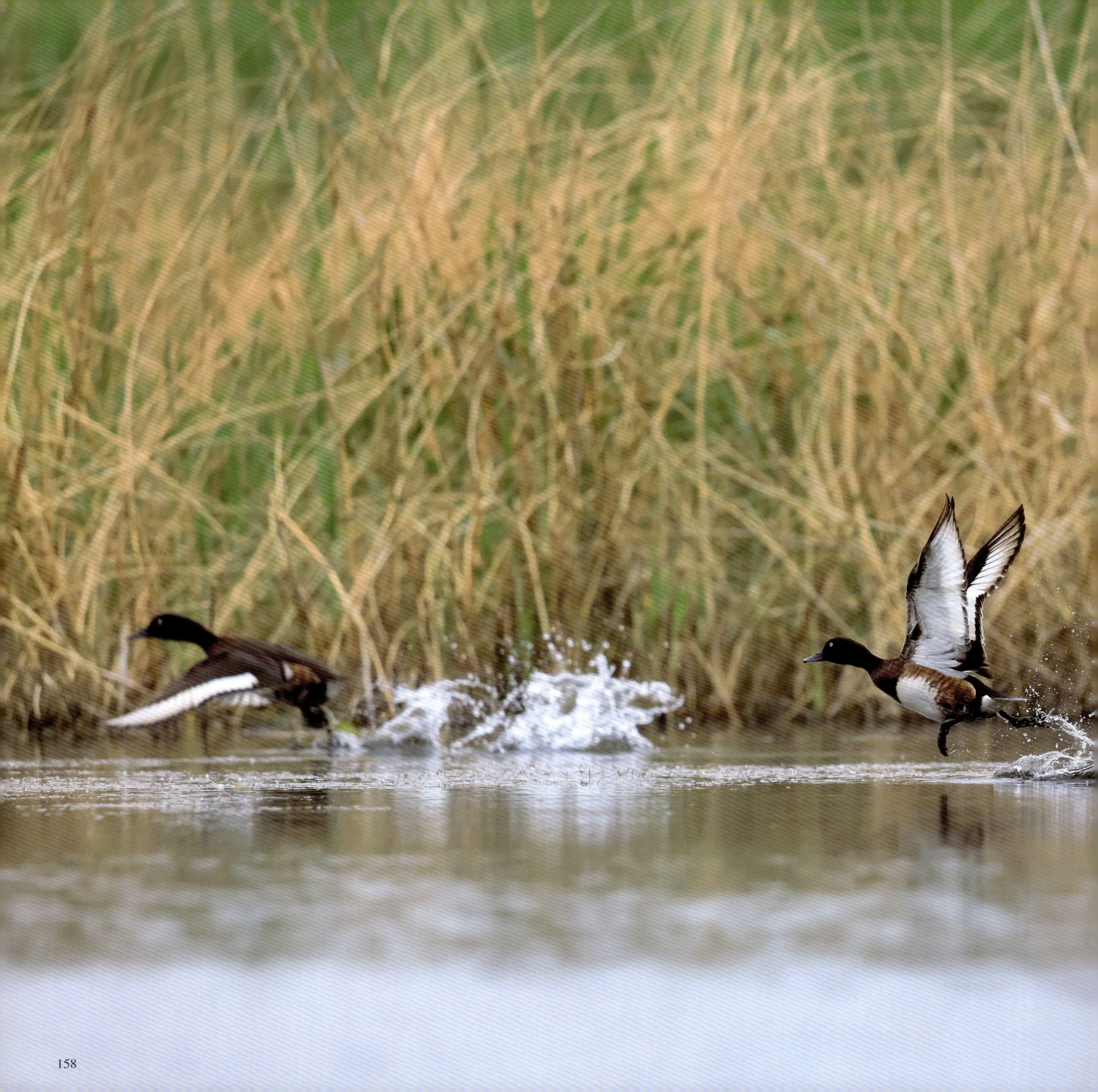

青头潜鸭

分布范围：繁殖于西伯利亚及中国东北地区，越冬于日本和朝鲜半岛、中国华南地区及东南亚等地。河北省内主要分布在衡水湖、白洋淀等湿地。

国家一级保护陆生野生动物，摄影：武明录

Baer's Pochar (*Aythya baeri*)

Distribution: It usually breeds in Siberia and northeast China, and overwinters in Japan and the Korean Peninsula, South China and Southeast Asia. Mainly distributed in Hengshui Lake, Baiyangdian and other wetlands in the Hebei province.

National first-class protected terrestrial wildlife

Photographed by Wu Minglu

东方白鹳

分布范围：在国外，通常繁殖于俄罗斯；在中国，基本繁殖于东北地区。河北省可见于沿海、衡水湖、白洋淀和平山湿地。
国家一级保护陆生野生动物，摄影：李克东

Oriental White Stork (*Ciconia boyciana*)

Distribution: Abroad, it usually breeds in Russia. In China, it basically breeds in the northeast. In Hebei Province, it found in coastal, Hengshui Lake, Baiyangdian and Pingshan wetland.
National first-class protected terrestrial wildlife
Photographed by Li Kedong

黑头白鹮

分布范围：分布于印度及东南亚地区，在我国繁殖于东北地区。在迁徙期，黑头白鹮会经河北、天津、山东、江苏、云南至福建、台湾、广东和海南等地越冬。在河北省秦皇岛偶见。
国家一级保护陆生野生动物，摄影：魏东

White Ibis (*Threskiornis melanocephalus*)

Distribution: It is distributed in India and Southeast Asia. It breeds in northeast China. During the migration period, they will through Hebei, Tianjin, Shandong, Jiangsu and Yunnan to overwinter in Fujian, Taiwan, Guangdong and Hainan. Occassionally found in Qinhuangdao, Hebei Province.
National first-class protected terrestrial wildlife
Photographed by Wei Dong

白鹤

分布范围：分布于中国、印度、阿富汗和日本等国。在河北省分布于沿海、坝上地区及衡水湖、白洋淀、文安、平山等地。在河北省为旅鸟。
国家一级保护陆生野生动物，摄影：刘庆顺

Siberian Crane (*Grus leucogeranus*)

Distribution: It is distributed in China, India, Afghanistan, Japan and other countries. In Hebei Province, it is distributed in coastal, Bashang areas and Hengshui Lake, Baiyangdian, Wen'an, Pingshan and other places, and as a passing migrant bird in there.
National first-class protected terrestrial wildlife
Photographed by Liu Qingshun

丹顶鹤

分布范围：分布于西伯利亚、蒙古国北部，于朝鲜、日本越冬。在中国的分布范围是东北地区及河北、北京、天津、山东、河南、内蒙古东部、长江中下游地区及东南沿海。在河北省是冬候鸟。
国家一级保护陆生野生动物，摄影：宗树兴

Red-crowned Crane (*Grus vipio*)
Distribution: It is distributed in Siberia, northern Mongolia, and overwinters in North Korea, Japan. In China, it is distributed in northeastern region and Hebei, Beijing, Tianjin, Shandong, Henan, eastern Inner Mongolia, middle and lower reaches of the Yangtze River and southeast coastal areas. In Hebei Province, it is a wintering migratory bird.
National first-class protected terrestrial wildlife
Photographed by Zong Shuxing

白枕鹤

分布范围：分布于俄罗斯、蒙古国、朝鲜和日本。在中国主要繁殖于黑龙江、吉林、内蒙古，越冬于江苏、安徽和江西。
国家一级保护陆生野生动物，摄影：刘建立

White-naped Crane (*Grus vipio*)
Distribution:It is found in Russia, Mongolia, North Korea and Japan. In China, it breeds mainly in Heilongjiang, Jilin and Inner Mongolia, and winters in Jiangsu, Anhui and Jiangxi.
National first-class protected terrestrial wildlife
Photographed by Liu Jianli

白头鹤

分布范围：分布于中国、日本、韩国、朝鲜、蒙古国、俄罗斯。河北省见于沿海及坝上地区。在河北省是冬候鸟。
国家一级保护陆生野生动物，摄影：李全江

Hooded Crane (*Grus monacha*)
Distribution:Distribution: distributed in China, Japan, South Korea, North Korea, Mongolia, Russia. Hebei Province is found in coastal and dam areas. In Hebei Province is a winter migratory bird.
National first-class protected terrestrial wildlife
Photographed by Li Quanjiang

栗斑腹鹀

分布范围：分布于我国的辽宁、内蒙古、河北等地区，河北省见于秦皇岛北戴河等地。
国家一级保护陆生野生动物，摄影：赵俊

Jankowski's Bunting (*Emberiza jankowskii*)
Distribution: It is distributed in Liaoning, Inner Mongolia, Hebei and other regions of China. In Hebei province, it is found in Beidaihe of Qinhuangdao city, and other places.
National first-class protected terrestrial wildlife
Photographed by Zhao Jun

■ 黄胸鹀

分布范围：在我国分布范围广泛——从东北到华南均有分布。此鸟在河北省为旅鸟。
国家一级保护陆生野生动物，摄影：张玉成

Yellow-breasted Bunting (*Embenzidae Emberiza aureola*)
Distribution: It is widely distributed in China, from northeast to South China. In Hebei Province, it is a passing migrant bird.
National first-class protected terrestrial wildlife
Photographed by Zhang Yucheng

小青脚鹬（左上图）

分布范围：繁殖于俄罗斯萨哈林岛（库页岛）、鄂霍次克海西岸，冬季迁徙途经日本、韩国及中国的东部河和石臼坨有记录。

国家一级保护陆生野生动物，摄影：周丰

Spotted Greenshank (*Tringa guttifer*) Top left

Distribution: It breeds in Sakhalin Island, of Russia, west coast of the Sea of Okhotsk, with records of winter migration via Japan, Korea and China East River and Shijiutuo.

National first-class protected terrestrial wildlife, Photographed by Zhou Feng

勺嘴鹬（左中图）

分布范围：繁殖于俄罗斯东北海岸，越冬于南亚、中南半岛。我国东部及东南沿海有分布。在河北省唐山曹妃甸、秦皇岛北戴河可见。

国家一级保护陆生野生动物，摄影：张玉成

Spoon-billed Sandpiper (*Calidris pygmeus*) Middle left

Distribution: It breeds in the northeast coast of Russia, overwintering in South Asia, Indochina Peninsula. It is distributed along the east and southeast coast of China. In Hebei Province, it can be seen in Tangshan Caofeidian and Qinhuangdao Beidaihe.

National first-class protected terrestrial wildlife, Photographed by Zhang Yucheng

黑鹳（下图）

分布范围：西班牙、非洲居多。中国主要分布于新疆。在河北省燕山及太行山地区，该鸟为繁殖鸟

国家一级保护陆生野生动物，摄影：聂立国。

Black Stork (*Ciconia nigra*) Figure below

Distribution: It is mostly distributed in Spain and Africa. In China, it is mostly found in Xinjiang. it is a breeding birds in Yanshan and Taihang Mountains of Hebei Province.

National first-class protected terrestrial wildlife, Photographed by Nie Liguo

■ **黑嘴鸥**（右上图）

分布范围：仅在中国东部沿海辽宁、河北、山东、江苏等地繁殖。
国家一级保护陆生野生动物，摄影：王秀荣

Saunders's Gull (*Saundersilarus saundersi*) Top right
Distribution: It breeds only on the east coast of China in Liaoning, Hebei, Shandong and Jiangsu.
National first-class protected terrestrial wildlife
Photographed by Wang Xiurong

■ **白肩雕**（右中图）

分布范围：在我国主要分布于新疆、甘肃等地区。河北省见于山区与丘陵。
国家一级保护陆生野生动物，摄影：郭勇敢

Asian Imperial Eagle (*Aquila heliaca*) Middle right
Distribution: In China, it is mainly distributed in Xinjiang, Gansu and other areas. In Hebei province, it is found in the mountains and hills.
National first-class pprotected terrestrial wildlife
Photographed by Guo Yongyong

■ **胡兀鹫**（下图）

分布范围：主要分布于亚洲、欧洲和非洲，在我国常见于西藏及附近地区，河北省内见于坝上与沿海。
国家一级保护陆生野生动物，摄影：刘庆顺

Bearded Vulture (*Gypaetus barbatus*) Figure below
Distribution: it is widespread throughout Asia, Europe and Africa. In China, it is usually found in Xizang and surrounding areas. In Hebei province, it is seen in Bashang Plateau and coastal areas.
National first-class protected terrestrial wildlife
Photographed by Liu Qingshun

彩鹮

分布范围：主要分布于欧洲、亚洲、非洲、美洲。在我国偶见于河北、上海、浙江、福建、台湾及广东沿海岛屿。在河北省偶见于衡水湖。

国家一级保护陆生野生动物，摄影：范琪

Glossy Ibis (*Plegadis falcinellus*)

Distribution: It is mainly distributed in Europe, Asia, Africa, and the Americas. In China, It is occasionally found in Hebei, Shanghai, Zhejiang, Fujian, Taiwan, and coastal islands of Guangdong. In Hebei Province, it is occasionally seen in Hengshui Lake.
National first-class protected terrestrial wildlife
Photographed by Fan Qi

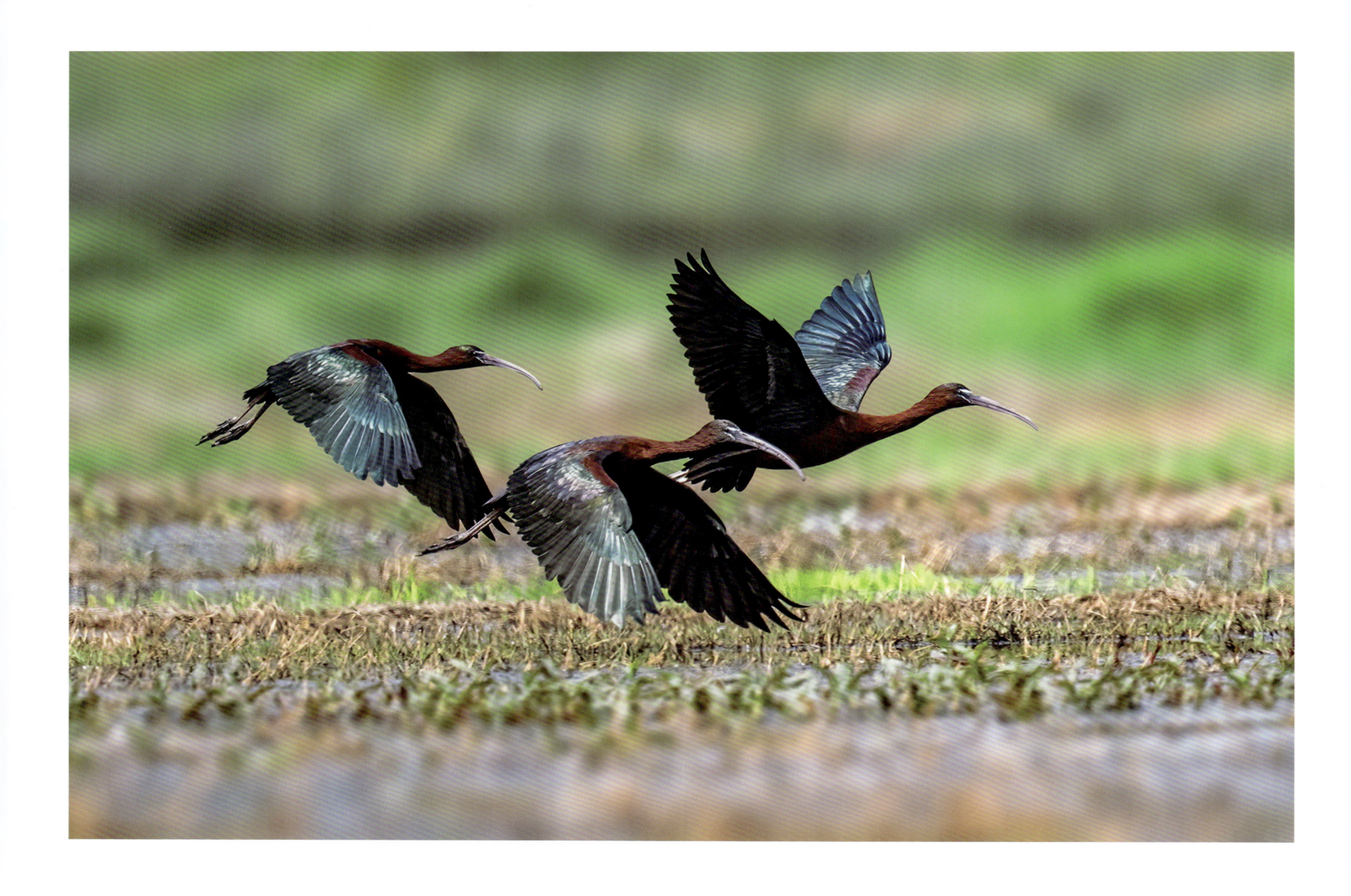

黑脸琵鹭

分布范围：迁徙时见于中国东北，在辽东半岛东侧的小岛上有繁殖记录。我国分布于东北、华北、华中与华南各省。
国家一级保护陆生野生动物，摄影：张玉成

Black-faced Spoonbill (*Platalea minor*)

Distribution: It is found in northeastern China during migration, with breeding records on small islands on the eastern side of the Liaodong Peninsula. In China, It is distributed in northeastern, northern, central and southern provinces.
National first-class protected terrestrial wildlife
Photographed by Zhang Yucheng

■ **黄嘴白鹭**

分布范围：主要分布于俄罗斯的远东地区、朝鲜、韩国和中国的东部地区。

国家一级保护陆生野生动物，摄影：黄冬青

Chinese Egret (*Egretta eulophotes*)

Distribution: It is mainly distributed in the Far East of Russian, North Korea, the Republic of Korea and the eastern part of China.

National first-class protected terrestrial wildlife

Photographed by Huang Dongqing

秃鹫

分布范围：在我国见于各省，在河北省为留鸟。
国家一级保护陆生野生动物，摄影：王秀荣

Cinereous Vulture (*Aegypius monachus*)
Distribution: In China, It is found in all provinces, and it is a resident bird in Hebei Province.
National first-class protected terrestrial wildlife
Photographed by Wang Xiurong

乌雕

分布范围：分布于欧洲东部，非洲东北部，亚洲东部、中部、南部和东南部等地。在我国河北省为旅鸟。
国家一级保护陆生野生动物，摄影：李克东

Greater Spotted Eagle (*Clanga clanga*)
Distribution: It is found in eastern Europe; northeastern Africa; eastern, central, southern and southeastern of Asia. In China, it is a passing migrant bird in Hebei Province.
National first-class protected terrestrial wildlife
Photographed by Li Kedong

草原雕

分布范围：分布于中国、阿尔巴尼亚、亚美尼亚、阿塞拜疆、巴林、孟加拉国、不丹、博茨瓦纳、保加利亚等许多国家。在我国河北省为夏候鸟。
国家一级保护陆生野生动物，摄影：王秀荣

Steppe Eagle (*Aquila nipalensis*)
Distribution: It is found in many countries such as China, Albania, Armenia, Azerbaijan, Bahrain, Bangladesh, Bhutan, Botswana and Bulgaria. It is a summering migratory bird in Hebei Province.
National first-class protected terrestrial wildlife
Photographed by Wang Xiurong

白尾海雕

分布范围：繁殖于欧亚大陆北部和格陵兰岛。在中国黑龙江省和内蒙古大兴安岭地区，白尾海雕为夏候鸟，其他地区为冬候鸟或旅鸟。在河北省多见于沿海地区，且为旅鸟。
国家一级保护陆生野生动物，摄影：王秀荣

White-tailed Sea-eagle (*Haliaeetus albicilla*)
Distribution: It breeds in northern Eurasia and Greenland. In China, it is a summering migratory bird in Heilongjiang Province and the Daxing'anling area of Inner Mongolia, and a wintering migratory bird or passing migrant bird in other areas. In Hebei Province, it is found in coastal areas, and it is a passing migrant bird in there.
National first-class protected terrestrial wildlife
Photographed by Wang Xiurong

虎头海雕

分布范围：分布于中国、日本、朝鲜、韩国和俄罗斯。在我国河北省见于沿海地区，为冬候鸟。
国家一级保护陆生野生动物，摄影：马宝祥

Stellers Sea Eagle (*Haliaeetus pelagicus*)
Distribution: It is distributed in China, Japan, North Korea, South Korea and Russia. It is seen in the coastal areas of Hebei Province in China and is a winter migratory bird.
National first-class protection of terrestrial wildlife
Photographed by Ma Baoxiang

猎隼

分布范围：分布于中欧、北非、印度北部、中亚至蒙古国及中国北方。在我国河北省为旅鸟。
国家一级保护陆生野生动物，摄影：赵永春

Saker Falcon (*Falco cherrug*)
Distribution: It is found in central Europe, north Africa, northern India, central Asia to Mongolia, and northern China. It is a passing migrant bird in Hebei Province.
National first-class protected terrestrial wildlife
Photographed by Zhao Yongchun

矛隼

分布范围：我国国内主要分布于黑龙江、吉林、辽宁、河北、内蒙古中部和新疆西部。在河北省，该鸟为冬候鸟。
国家一级保护陆生野生动物，摄影：徐捷

Gyrfalcon (*Falco rusticolus*)
Distribution: In China, it is distributed in Heilongjiang, Jilin, Liaoning, Hebei, central Inner Mongolia and western Xinjiang. It is a wintering migratory bird in Hebei Province.
National first-class protected terrestrial wildlife
Photographed by Xu Jie

金雕

分布范围：分布于北半球温带、亚寒带和寒带地区。中国多见于北方地区。河北省多见于丘陵与山区，且为留鸟或旅鸟。
国家一级保护陆生野生动物，摄影：王秀荣

Golden Eagle (*Aquila chrysaetos*)
Distribution: It is distributed in the temperate, subfrigid and frigid regions of the northern hemisphere. In China, it is mostly found in the northern part. In Hebei Province, it is mostly found as a resident bird or passing migrant bird in hilly and mountainous areas.
National first-class protected terrestrial wildlife
Photographed by Wang Xiurong

玉带海雕

分布范围：主要分布于里海和黄海之间的亚欧大陆地区——从哈萨克斯坦到蒙古，从喜马拉雅山脉到印度北部的亚洲中部地区。在我国河北省见于沿海与坝上地区。该鸟在河北省为旅鸟。

国家一级保护陆生野生动物，摄影：高云江

Pallas's Fish Eagle (*Haliaeetus leucoryphus*)

Distribution: Mainly found in the Eurasian region between the Caspian Sea and the Yellow Sea – from Kazakhstan to Mongolia, from the Himalayas to northern India in central Asia. In Hebei Province of China, it is found in coastal and dam areas. The bird is a traveling bird in Hebei Province.

National first-class protected terrestrial wildlife

Photographed by Gao Yunjiang

自然保护区资源

自然保护区在保护生物多样性和维护生态平衡、实现人与自然和谐相处等方面发挥了重要的生态平衡作用。每一处自然保护区都是大自然的杰作，见证了时间的变迁和生命的奇迹。

河北省有13个国家级自然保护区，包括以保护森林和其他植被为主的保护区、以保护珍贵稀有野生动物为主的保护区和以保护自然历史遗迹为主的保护区，具有重要的生态、保护、科研价值。

河北昌黎黄金海岸国家级自然保护区位于秦皇岛市昌黎县，为海洋类型自然保护区。这里的主要保护对象是沙丘、沙堤、潟湖、林带和海洋生物等构成的沙质海岸自然景观和沿岸海洋生态系统。

河北围场红松洼国家级自然保护区位于承德市围场满族蒙古族自治县的最北部。这里以塞罕坝曼甸山地草甸生态系统及珍稀野生动植物多样性和滦河、西辽河河源湿地景观生境为主要保护对象。

河北雾灵山国家级自然保护区坐落于燕山山脉南缘，位于承德市兴隆县，其主要保护对象为温带森林生态系统和猕猴。

驼梁国家级自然保护区位于石家庄市平山县，主要保护对象为森林生态系统和珍稀濒危野生动植物物种。

青崖寨国家级自然保护区位于邯郸武安市境内，在极小种群物种保护方面具有重要意义。

河北塞罕坝国家级自然保护区位于承德市西部，是森林—草原交错带生态系统，有着"中国绿色明珠"和"华北绿宝石"的美誉。

河北小五台山国家级自然保护区地处太行山山脉中段的小五台山区。这里主要保护的对象是暖温带森林生态系统和褐马鸡等国家重点保护野生动植物。

河北衡水湖国家级自然保护区位于衡水市区，拥有华北地区最大的内陆淡水湖之一，其主要保护对象为内陆淡水湿地生态系统和国家一、二级保护鸟类。

河北茅荆坝国家级自然保护区位于承德市隆化县，是森林生态系统及其生物多样性、珍稀濒危物种及其栖息地、自然生态环境和滦河上游的水源地。

河北滦河上游国家级自然保护区位于承德市围场满族蒙古族自治县境内。其主要保护对象为滦河上游的自然生态环境、森林生态系统以及生物多样性和珍稀濒危的野生动植物物种。

河北大海陀国家级自然保护区位于张家口市赤城县。其主要保护对象为暖温带森林生态系统类型及珍稀濒危野生动植物物种。

河北泥河湾国家级自然保护区位于张家口市阳原县东部。其主要保护对象为晚新生代典型地层剖面、新生代地层中的哺乳动物化石及其主要发掘遗址、地层中的人类文化遗迹和古人类活动遗址等。这里被称为"研究远古人类的百科全书"。

河北柳江盆地地质遗迹国家级自然保护区位于秦皇岛市北部。这里有全国唯一一个地质变化最全的地质遗迹。该保护区主要的保护对象为标准地质剖面、典型地质构造等地质遗迹，这里被地学界誉为"地学百科全书"。

Nature Reserve Resources

Nature reserves play an important role in protecting biodiversity, maintaining ecological balance, and realizing the harmonious coexistence between man and nature. Each nature reserve is a masterpiece of nature, a testament to the vicissitudes of time and the miracle of life.

There are 13 national nature reserves in Hebei Province, including the protected areas for the protection of forests and other vegetation, the protected areas for the protection of precious and rare wild animals, and the protected areas for the protection of natural and historical relics, which have important ecological, conservation and scientific research values.

Hebei Changli Gold Coast National Nature Reserve is located in Changli County, Qinhuangdao City, is a marine nature reserve. The main objects of protection are the sandy coastal natural landscape and coastal marine ecosystem composed of sand dunes, sand banks, lagoons, forest belts and marine life.

Hebei Weichang Hongsongwa National Nature Reserve is located in the northernmost part of Weichang Manchu and Mongolian Autonomous County in Chengde City. The main protection objects are the grassland ecosystem and rare wildlife diversity in the Mandian mountain meadow of Saihanba and the wetland landscape habitat of the headwaters of the Luanhe River and the Xiliaohe River.

Hebei Wuling Mountain National Nature Reserve is located at the southern edge of the Yanshan Mountains. It is located in Xinglong County, Chengde City, and is mainly protected object is temperate forest ecosystems and macaques.

Tuoliang National Nature Reserve is located in Pingshan County, Shijiazhuang City. There mainly protects forest ecosystems and rare and endangered wild animal and plant species.

Qingyazhai National Nature Reserve is located in Wu'an City, Handan, and is of great significance in the conservation of species with very small populations.

Saihanba National Nature Reserve in Hebei Province is located in the west of Chengde City. It is an ecosystem in the forest-grassland ecotone, and has the reputation of "China's Green Pearl" and "North China's Emerald".

Hebei Xiaowutai Mountain National Nature Reserve is located in the Xiaowutai Mountains in the middle of the Taihang Mountains. The main objects of protection are warm temperate forest ecosystems and wild animals and plants under national key protection, such as brown pheasants.

Hebei Hengshui Lake National Nature Reserve is located in Hengshui City. There is one of the largest inland freshwater lakes in North China, and the main objects of protection are inland freshwater wetland ecosystems and national first and second class birds.

Maojingba National Nature Reserve is located in Longhua County, Chengde City. There is a forest ecosystem and its biodiversity, rare and endangered species and their habitats, natural ecological environment and water source of the upper reaches of the Luanhe River.

Hebei Luanhe River Upper Reaches National Nature Reserve is located in Weichang Manchu Mongolian Autonomous County, Chengde City. The main objects of protection are the natural ecological environment, forest ecosystem, biodiversity and rare and endangered wild animal and plant species in the upper reaches of the Luanhe River.

Hebei Dahaituo National Nature Reserve is located in Chicheng County, Jiakou City, Chengde City. The main objects of protection are warm temperate forest ecosystem types and rare and endangered species of wild animals and plants.

Nihewan National Nature Reserve in Hebei Province is located in the east of Yangyuan County, Zhangjiakou City. The main objects of protection are the typical stratigraphic sections of the Cenozoic, the mammalian fossils in the Late Cenozoic strata and their main excavation sites, the human cultural relics in the strata and the sites of ancient human activities, etc. There are known as "encyclopedia for the study of ancient humans".

Hebei Liujiang Basin Geological Relics National Nature Reserve is located in the north of Qinhuangdao City. There is the only geological relics with the most complete geological changes in the country, and the main protection objects are the geological relics such as standard geological sections and typical geological structures, which are praised as "encyclopedia of geology" by the geological field.

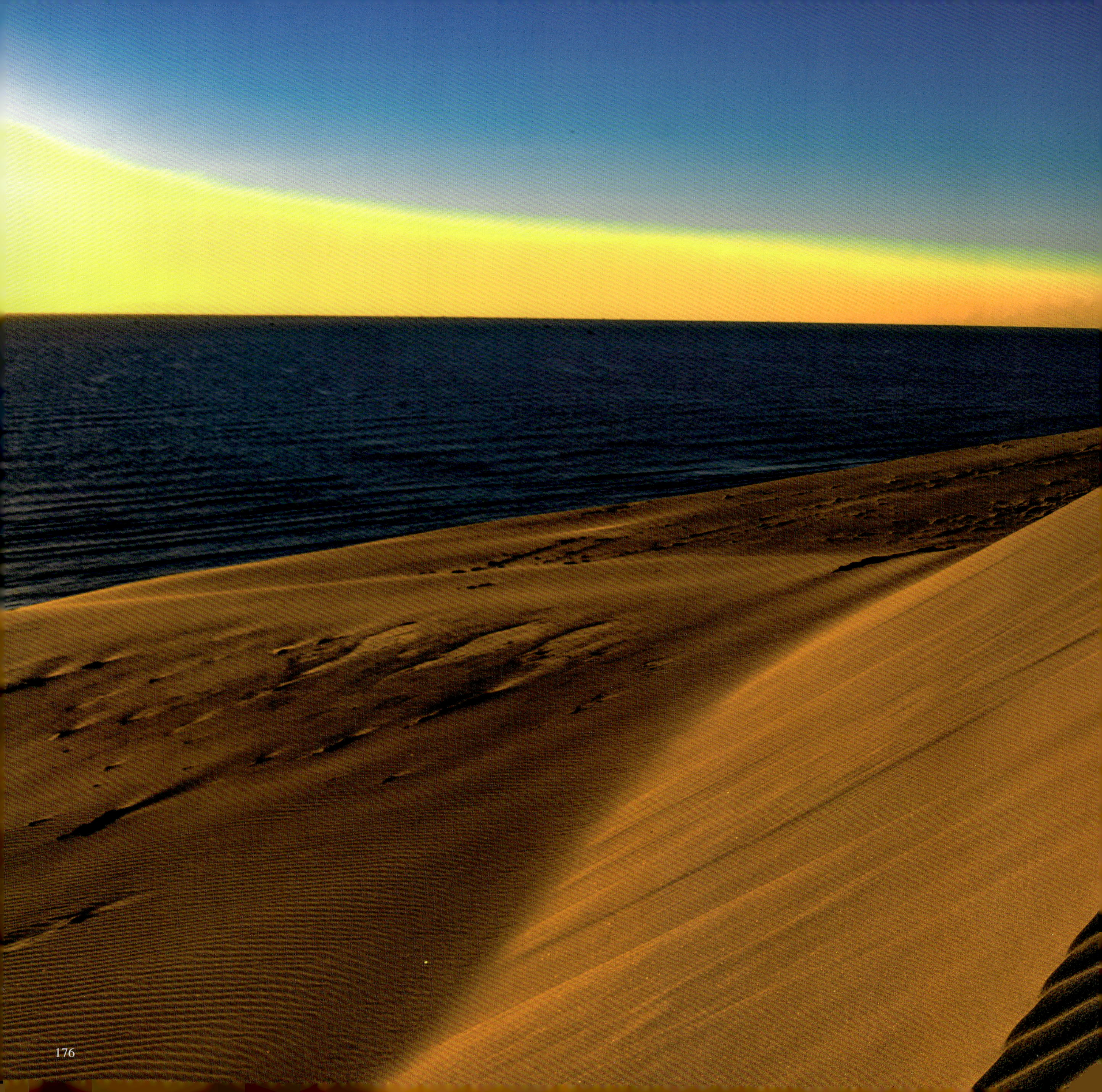

河北昌黎黄金海岸国家级自然保护区

　　河北昌黎黄金海岸国家级自然保护区位于河北省东北部秦皇岛市昌黎县沿海，这里的主要保护对象为沙丘、沙堤、潟湖、林带和海洋生物等构成的沙质海岸自然景观及所在海区的生态环境和自然资源。

《河北昌黎黄金海岸国家级自然保护区》，秦皇岛市昌黎县，摄影：费长波

Changli Golden Coast National Nature Reserve, Hebei Province
It is located along the coast of Changli County, Qinhuangdao City, in the northeastern part of Hebei Province. The main protected objects of the reserve are the sandy coastal natural landscapes, which include sand dunes, sandbanks, lagoons, forest belts, and marine organisms, as well as the ecological environment and natural resources of the surrounding sea area.
Hebei Changli Gold Coast National Nature Reserve, Changli County, Qinhuangdao City
Photographed by Fei Changbo

河北围场红松洼国家级自然保护区

　　河北围场红松洼国家级自然保护区位于河北省围场满族蒙古族自治县境内，是一个以塞罕坝曼甸山地草甸生态系统、珍稀野生动植物多样性及滦河、西辽河河源湿地景观生境为主要保护对象的综合性草地类自然保护区。
《河北围场红松洼国家级自然保护区》，承德市围场县，摄影：林树国

Weichang Hongsongwa National Nature Reserve, Hebei Province
It is located within the jurisdiction of Weichang Manchu and Mongolian Autonomous County in Hebei Province. There is a comprehensive grassland nature reserve primarily focused on protecting the Saihanba meadow mountain ecosystem, the diversity of rare wildlife and plants, as well as the wetland landscape habitats of the sources of the Luan River and the West Liaohe River.
Hebei Weichang Hongsongwa National Nature Reserve, Weichang County, Chengde City
Photographed by Lin Shuguo

河北雾灵山国家级自然保护区

河北雾灵山国家级自然保护区位于河北省兴隆县北部，这里珍藏着较完整的森林生态系统。据统计，这里有高等植物 165 科 645 属 1870 种，其中苔藓植物 47 科 128 属 317 种，蕨类植物 15 科 24 属 65 种，裸子植物 2 科 6 属 13 种，被子植物 104 科 507 属 1475 种。
《她在丛中笑》，承德市雾灵山主峰，摄影：张希军

Wuling Mountain National Nature Reserve, Hebei Province
located in the northern part of Xinglong County, Hebei Province, the Wuling Mountain Nature Reserve houses a relatively complete forest ecosystem. According to statistics, there are 1870 species of higher plants belonging to 645 genera in 165 families, including 317 species of bryophytes belonging to 128 genera in 47 families, 65 species of ferns belonging to 24 genera in 15 families, 13 species of gymnosperms belonging to 6 genera in 2 families, and 1475 species of angiosperms belonging to 507 genera in 104 families.
She Laughs in the Bushes, The main peak of Wuling Mountain in Chengde City
Photographed by Zhang Xijun

河北驼梁国家级自然保护区

　　河北驼梁国家级自然保护区位于河北省平山县西北部太行山中段山区，主要保护对象为森林生态系统和珍稀濒危野生动植物物种。
《驼梁风光》，保定市阜平县驼梁风景区，摄影：杨丽影

Tuoliang National Nature Reserve, Hebei Province
It is located in the northwestern part of Pingshan County, Hebei Province, within the central section of the Taihang Mountains. Its protected objects are the forest ecosystems and rare and endangered species of wild fauna and flora.
Camel Liang Scenery, Camel Liang Scenic Area, Fuping County, Baoding City
Photographed by Yang Liying

河北青崖寨国家级自然保护区

河北青崖寨国家级自然保护区位于河北省武安市西北部的太行深山区。这里的主要保护对象为森林生态系统、珍稀濒危野生动植物物种及地质遗迹，这里属于森林生态系统类型自然保护区。

《青崖寨》，武安市，摄影：李树锋

Qingyazhai National Nature Reserve, Hebei Province

It is located in the deep mountainous area of the Taihang Mountains in the northwestern part of Wu'an City, Hebei Province. The main protected objects are the forest ecosystem, rare and endangered wild animal and plant species, and geological relics. It belongs to the nature reserve of the forest ecosystem type.

Qingya Village, Wu'an City
Photographed by Li Shufeng

河北塞罕坝国家级自然保护区

河北塞罕坝国家级自然保护区位于河北省承德市围场满族蒙古族自治县境内，其主要保护对象是森林—草原交错带生态系统，滦河、辽河水源地，黑鹳、金雕等珍稀濒危动植物物种。这里属于森林生态系统类型自然保护区。
《河北塞罕坝国家级自然保护》，承德市围场县，摄影：林树国

Saihanba National Nature Reserve, Hebei Province
It is located within Weichang Manchu and Mongolian Autonomous County in Chengde City, Hebei Province. Its primary protected objects are the forest-grassland ecotone ecosystem, water sources of Luanhe and Liaohe rivers, and rare and endangered species such as the black stork and golden eagle. It is a nature reserve of the forest ecosystem type.
Hebei Saihanba National Nature Conservation, Weichang County, Chengde City
Photographed by Lin Shuguo

河北小五台山国家级自然保护区

小五台山国家级自然保护区位于河北省西北张家口地区的蔚县和涿鹿两县境内。该保护区类型属于森林野生动物类型自然保护区，主要保护对象是天然针阔混交林、亚高山灌丛、草甸、国家一级重点保护动物褐马鸡。

《河北小五台山国家级自然保护区》，张家口市地区的蔚县和涿鹿两县境内，摄影：赵全胜

Xiaowutai Mountain National Nature Reserve, Hebei Province
It is located within the territories of Yuxian County and Zhuolu County in the Zhangjiakou region of northwest Hebei Province. This reserve is classified as a forest and wildlife nature reserve. Its primary protected objects are includes natural coniferous and broad-leaved mixed forests, subalpine shrublands, meadows, and the Brown Eared Pheasant, which is a first-class state-protected wild animal in China.

Hebei Xiaowutai Mountain National Nature Reserve, In the territory of Wei County and Zhuolu County in the area of Jiakou City
Photographed by Zhao Quansheng

河北衡水湖国家级自然保护区

　　河北衡水湖国家级自然保护区坐落于河北省衡水市桃城区和冀州两县区境内。河北衡水湖国家级自然保护区生物多样性十分丰富，以内陆淡水湿地生态系统和国家一、二级保护鸟类为主要保护对象，属淡水湿地生态系统类型自然保护区。

《夏日衡水湖》，衡水市，摄影：康同跃

Hengshui Lake National Nature Reserve, Hebei Province

Situated within the territories of Taocheng District and Jizhou County in Hengshui City, Hebei Province, the Hengshui Lake National Nature Reserve is rich in biodiversity. Its primary protected objects are on the inland freshwater wetland ecosystem and national first and second class protected bird species. It is classified as a freshwater wetland ecosystem type nature reserve.

Summer Hengshui Lake, Hengshui City
Photographed by Kang Tongyue

河北大海陀国家级自然保护区

　　大海陀国家级自然保护区位于首都北京西北、河北省赤城县西南。该保护区是典型的山地森林生态系统类型，在华北地区的植被垂直地带性和生物地理区系等方面具有典型性和代表性。

《河北大海坨国家级自然保护区》，张家口市赤城县西南，供图：视觉中国

Dahaituo National Nature Reserve, Hebei Province
located to the northwest of Beijing and in the southwest of Chicheng County, Hebei Province, this reserve is a typical mountain forest ecosystem type. It has significant representativeness in terms of vegetation vertical zonation and biogeographic regions in North China.
Hebei Dahaituo National Nature Reserve, Southwest of Chicheng County, Zhangjiakou City
Photo provided by Visual China

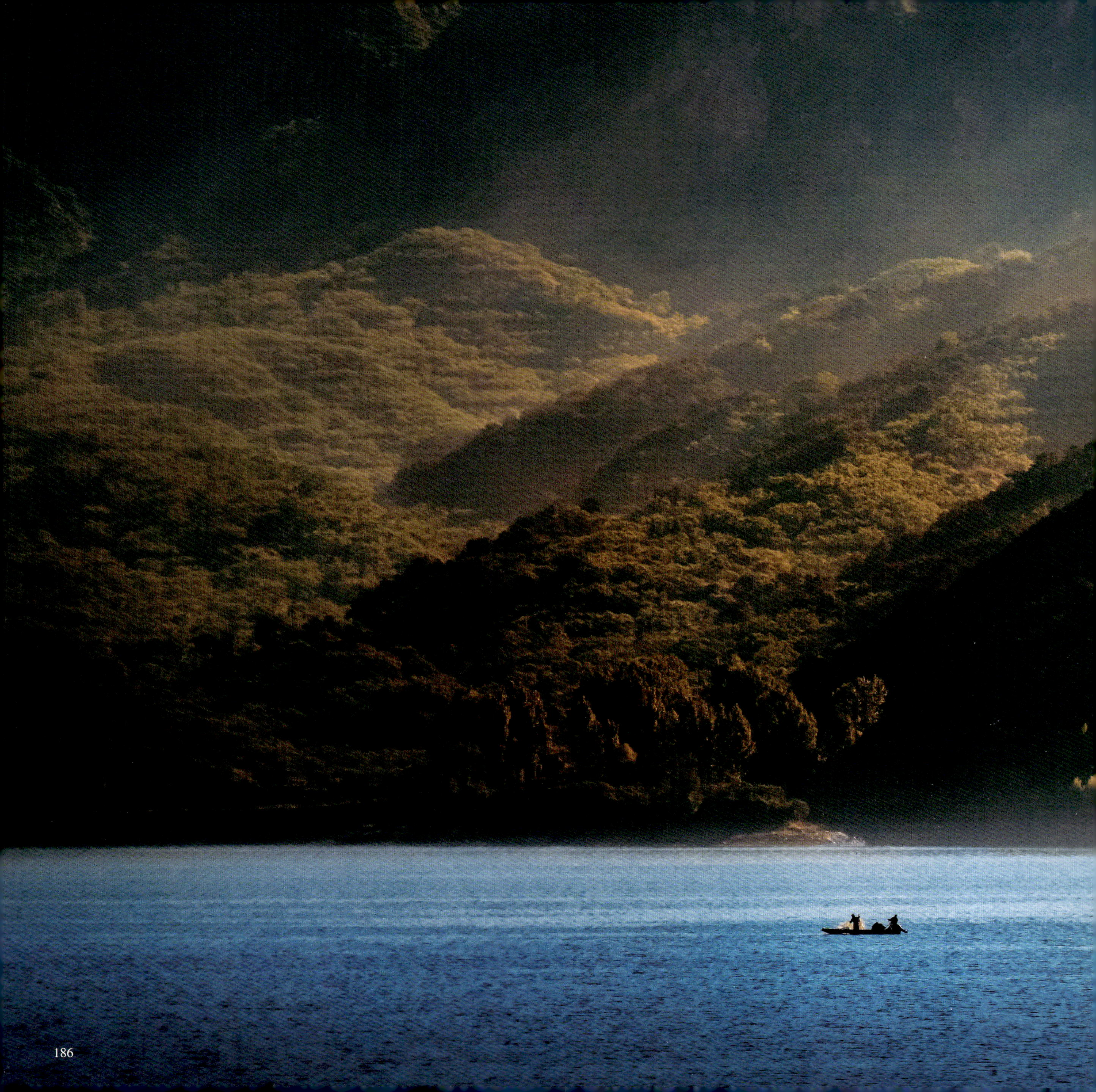

河北滦河上游国家级自然保护区

  滦河上游国家级自然保护区位于河北省承德市围场满族蒙古族自治县境内的接坝地区。滦河上游国家级自然保护区的主要保护对象为滦河上游的自然生态环境，森林、草原生态系统及其生物多样性和珍稀濒危的野生动植物物种。这里是融典型森林生态系统及野生动植物保护于一身的生态系统类大型自然保护区。

《滦河风光》，承德市围场县，摄影：薛志军

Luanhe River Upper Reaches National Nature Reserve, Hebei Province
Located in the Jieba area within the territory of Weichang Manchu and Mongolian Autonomous County, Chengde City, Hebei Province, Luanhe River Upper Reaches National Nature Reserve mainly focuses on protecting the natural ecological environment of the upper reaches of the Luanhe River, the forest and grassland ecosystems, as well as their biodiversity and rare and endangered wild animal and plant species. There is a large-scale ecosystem-type nature reserve that integrates typical forest ecosystems and wildlife conservation.
Luanhe River Scenery, Weichang County, Chengde City
Photographed by Xue Zhijun

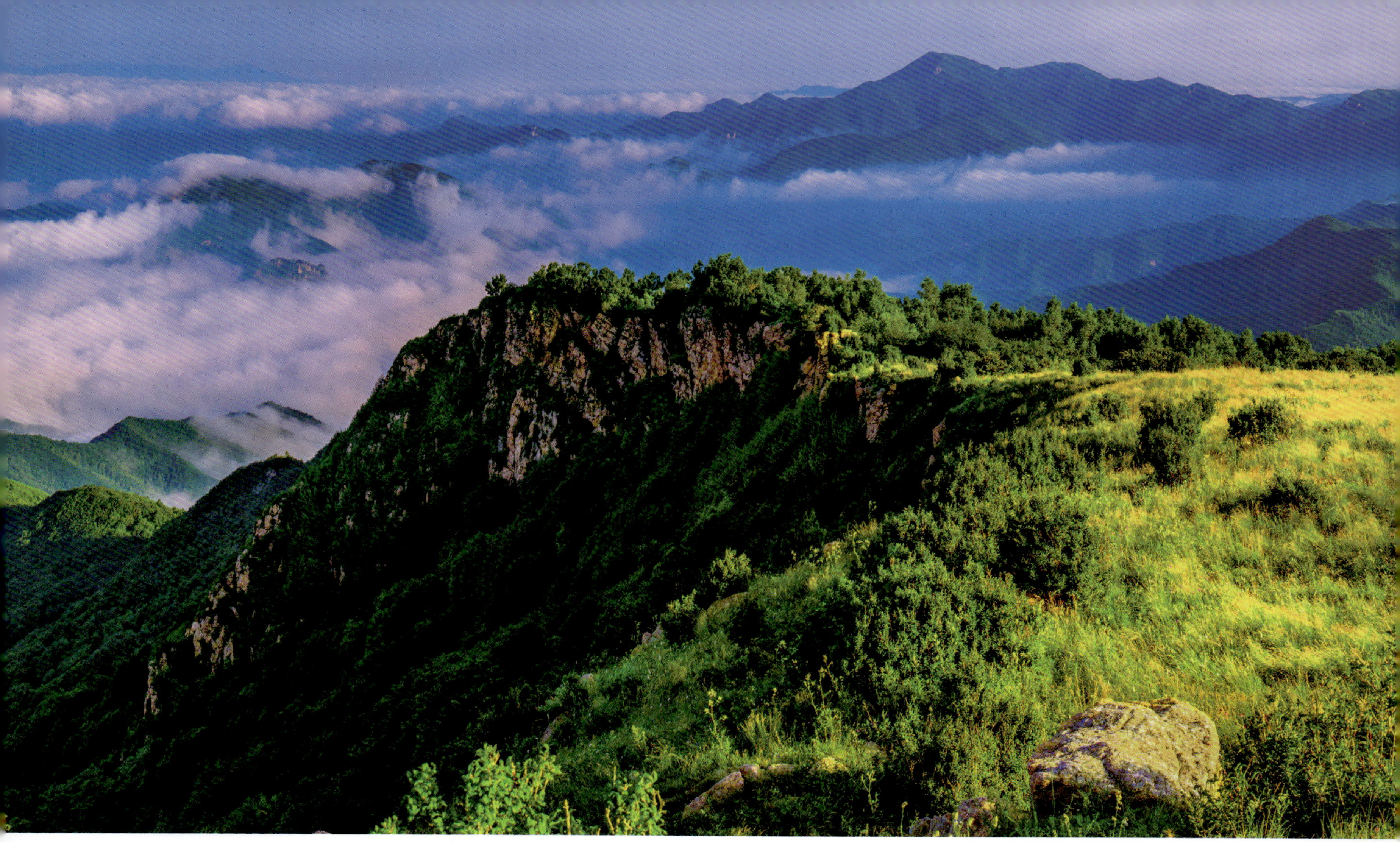

河北茅荆坝国家级自然保护区

　　河北茅荆坝国家级自然保护区位于河北省承德市隆化县境内。该保护区属于森林生态系统类型的自然保护区。

《远眺茅荆坝林海》，承德市隆化县茅荆坝国有林场，摄影：张志栋

Maojingba National Nature Reserve, Hebei Province
It is located within the territory of Longhua County, Chengde City, Hebei Province. The reserve is classified as a forest ecosystem type nature reserve.
Overlooking the Maojingba forest sea, Maojingba State-owned Forest Farm in Longhua County, Chengde City
Photographed by Zhang Zhidong

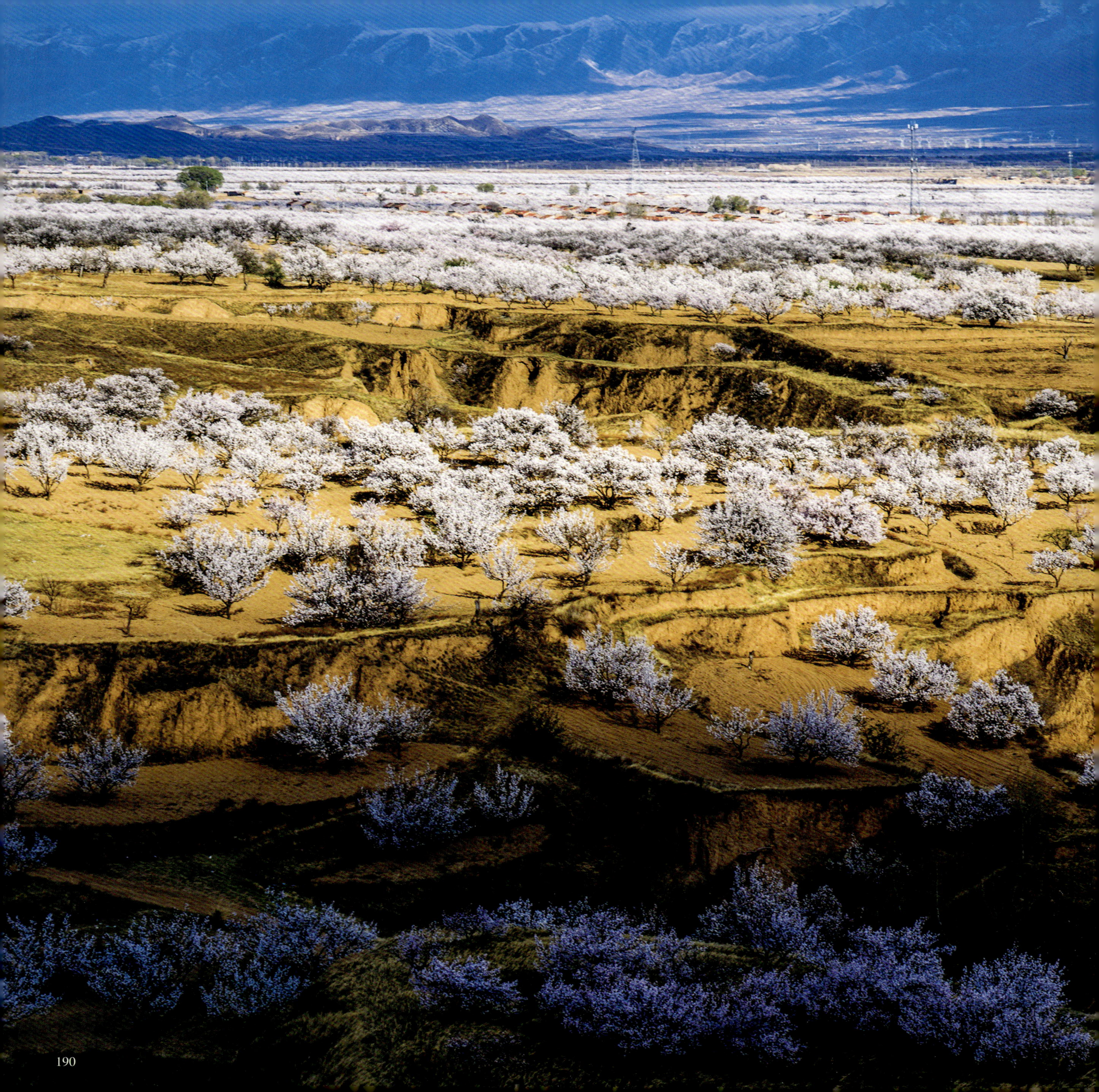

河北泥河湾国家级自然保护区

泥河湾国家级自然保护区位于河北省阳原县、蔚县以及山西省的雁北地区。其主要保护对象为晚新生代典型地层剖面、晚新生代地层中的哺乳动物化石及其主要发掘遗址、地层中的人类文化遗迹和古人类活动遗址等。

《泥河湾春色》，张家口市蔚县，摄影：潘辉峰

Nihewan National Nature Reserve, Hebei Province
It is located in Yangyuan County and Yuxian County of Hebei Province, as well as the Yanbei area of Shanxi Province. The main protected objects are the typical stratigraphic sections of the Late Cenozoic Era, the mammalian fossils in the Late Cenozoic strata and their main excavation sites, human cultural relics in the strata and the sites of ancient human activities.
Spring in Mud River Bay, Wei County, Zhangjiakou City
Photographed by Pan Huifeng

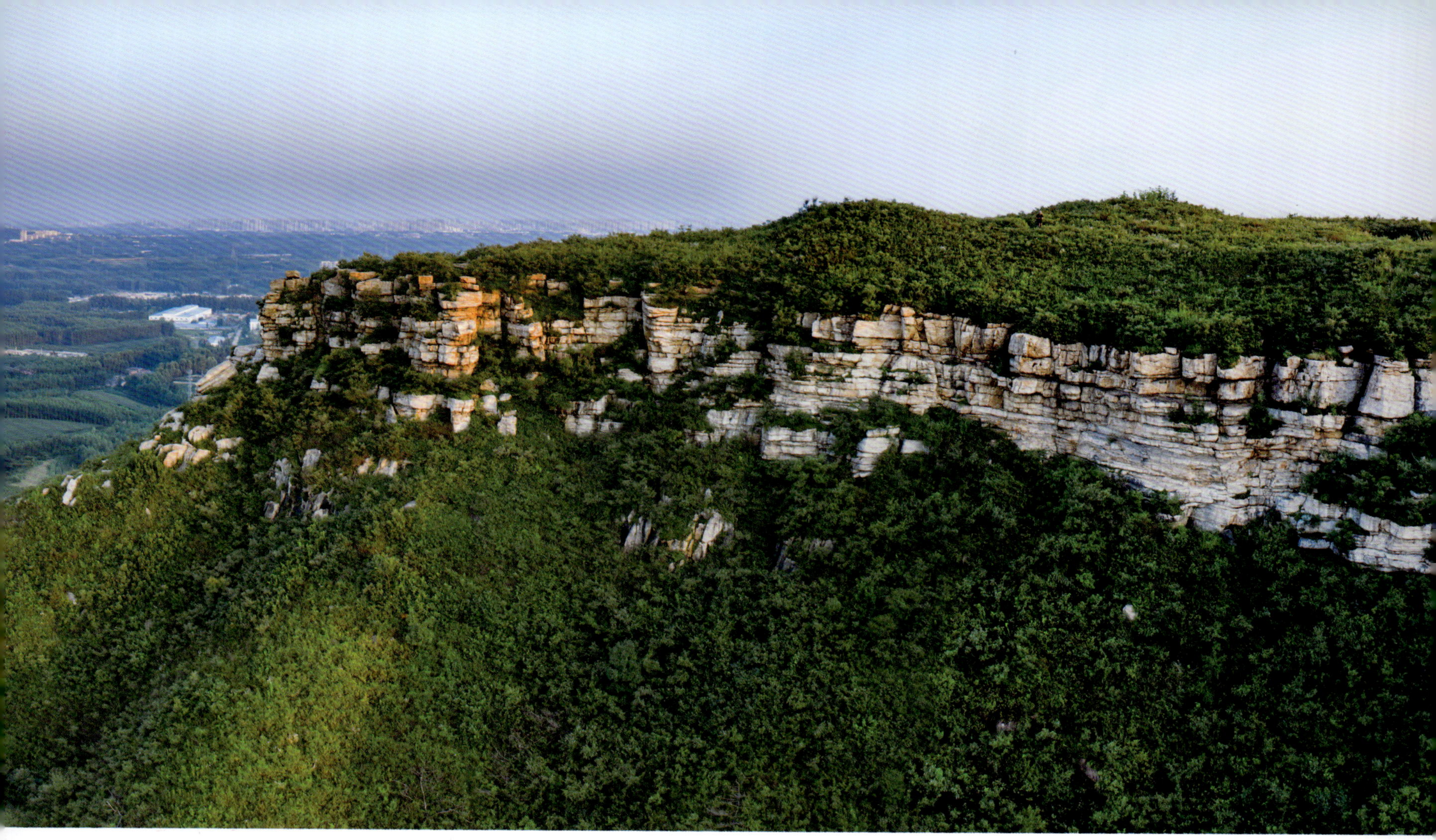

河北柳江盆地地质遗迹国家级自然保护区

河北柳江盆地地质遗迹国家级自然保护区位于秦皇岛市以北的燕山山脉东段与华北平原接壤的区域。柳江盆地蕴藏着极其丰富的地质遗迹，集典型性、自然性、稀有性、多样性、系统性和完整性于一体，享有"天然实验室"和"自然博物馆"的美誉。

《鸡冠山地堑》，秦皇岛市，摄影：徐树春

Liujiang Basin Geological Relics National Nature Reserve, Hebei Province
It is located in the area where the eastern section of the Yanshan Mountains to the north of Qinhuangdao City meets the North China Plain. The Liujiang Basin is rich in geological relics, characterized by their typicality, naturalness, rarity, diversity, systematicness and completeness, and is renowned as "a natural laboratory" and "a natural museum".
Jiguan Mountain Graben, Qinhuangdao City
Photographed by Xu Shuchun.

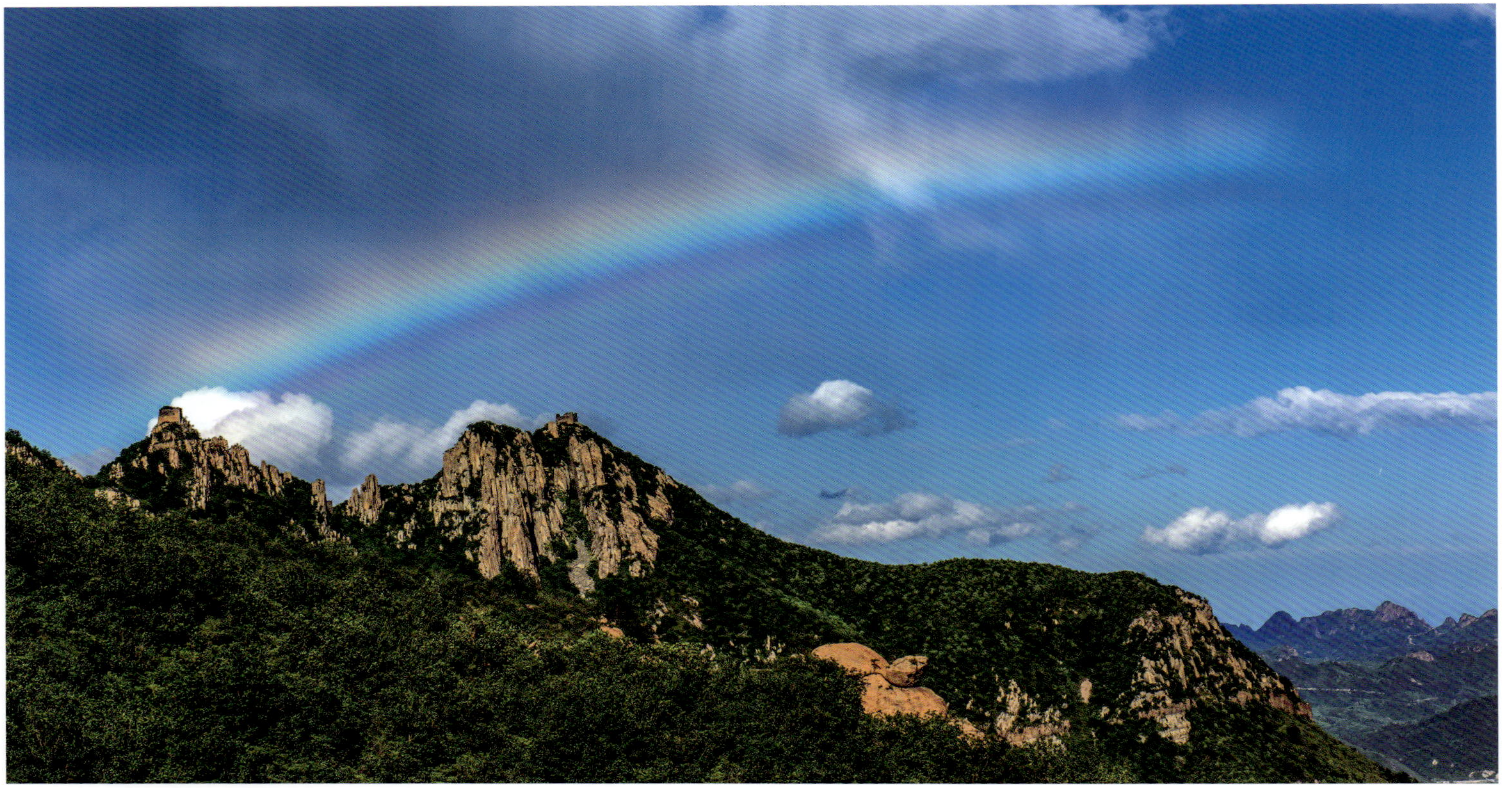

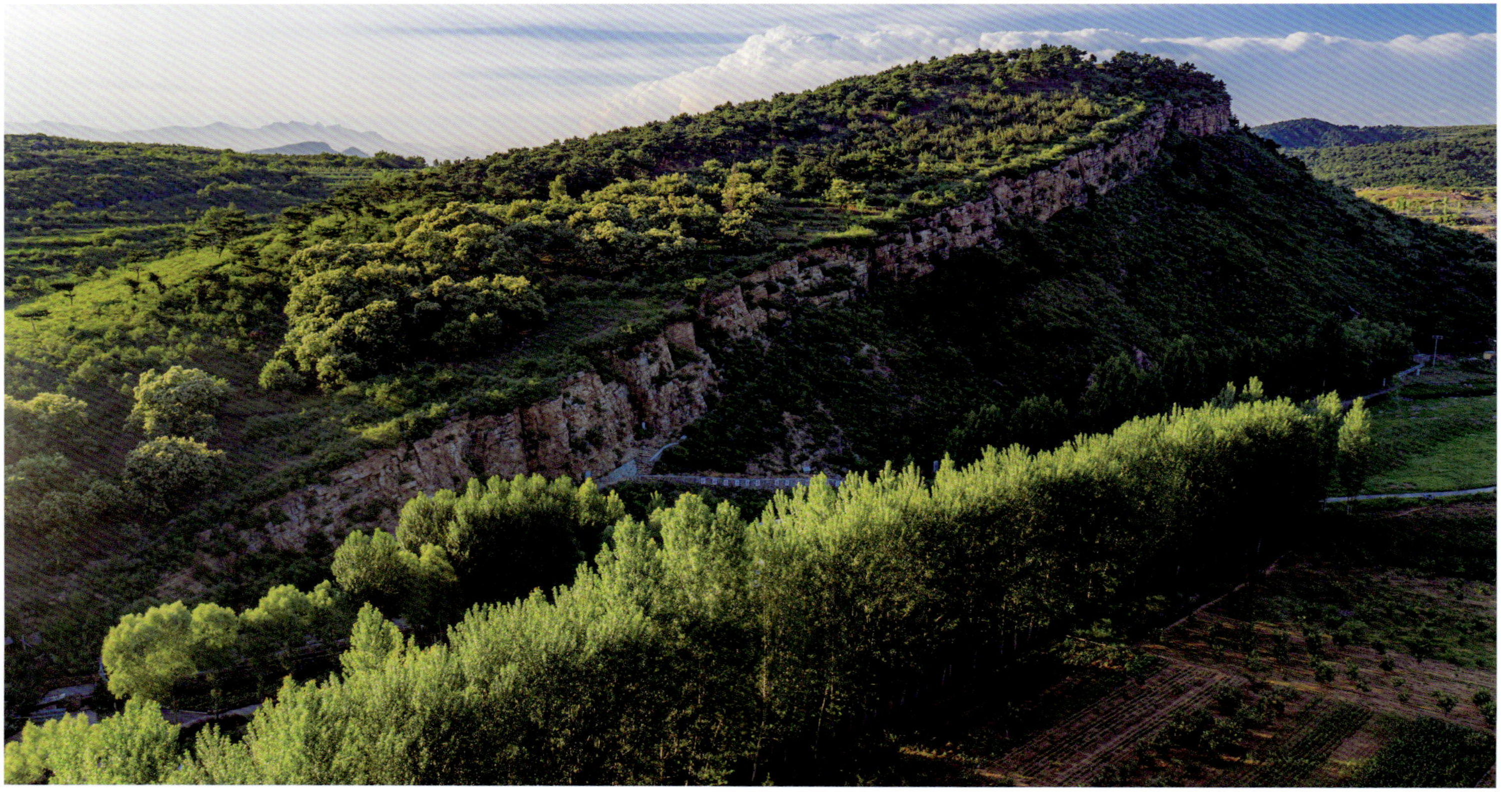

河北柳江盆地地质遗迹国家级自然保护区

Liujiang Basin Geological Relics National Nature Reserve, Hebei Province

《柳江盆地板厂峪地貌》，秦皇岛市海港区板厂峪长城，摄影：田利君（对页上图）
Landforms of Bancangyu in Liujiang Basin, Qinhuangdao City Yu Great Wall
Photographed by Tian Lijun (Above of Opposite Page)

《张岩子村西倾斜的石岩》，秦皇岛市海港区驻操营镇，摄影：徐树春（对页下图）
The Rocks Sloping Westward in Zhangyanzi Village, Haigang District, Qinhuangdao City
Photographed by Xu Shuchun (Bottom of Opposite Page)

《侏罗系髫髻山组地层》，秦皇岛市海港区板厂峪长城，摄影：崔重辉（上左图）
Jurassic Tiaoji Mountain Formation Stratigraphy, Qinhuangdao City Yu Great Wall
Photographed by Cui Zhonghui (Top Left)

《踏岩远眺，断崖为龙山组砂岩》，秦皇岛市海港区板厂峪长城，摄影：崔重辉（上右图）
Gazing from the Rocks, with the Clifftop of Longshan Formation Sandstone, Qinhuangdao City Yu Great Wall
Photographed by Cui Chonghui (Top Right)

《柳江盆地全貌》，秦皇岛市海港区石门寨镇，摄影：崔重辉（下图）
Overall View of Liujiang Basin, Shimenzhai Town, Haigang District, Qinhuangdao City
Photographed by Cui Zhonghui (Below)

■ 河北柳江盆地地质遗迹国家级自然保护区
Liujiang Basin Geological Relics National Nature Reserve, Hebei Province

《山羊寨奥陶系地层构造运动导致地层直立》，秦皇岛市海港区，摄影：崔重辉（上图）
Ordovician System in Shanyangzhai Resulting in Upright Strata, Haigang District, Qinhuangdao City Photographed by Cui Zhonghui (Top)

《山羊寨奥陶系冶里组地层》，秦皇岛市海港区，摄影：崔重辉（下图）
Ordovician Yeli Formation Stratigraphy in Shanyangzhai, Haigang District, Qinhuangdao City Photographed by Cui Zhonghui (Bottom)

《鸡冠山断崖——上元古界青白口系龙山组砂岩》，秦皇岛市海港区，摄影：崔重辉（右图）
Jiguan Mountain Cliff - Upper Proterozoic Qingbaikou System Longshan Formation Sandstone, Haigang District, Qinhuangdao City, Photographed by Cui Zhonghui (Right)

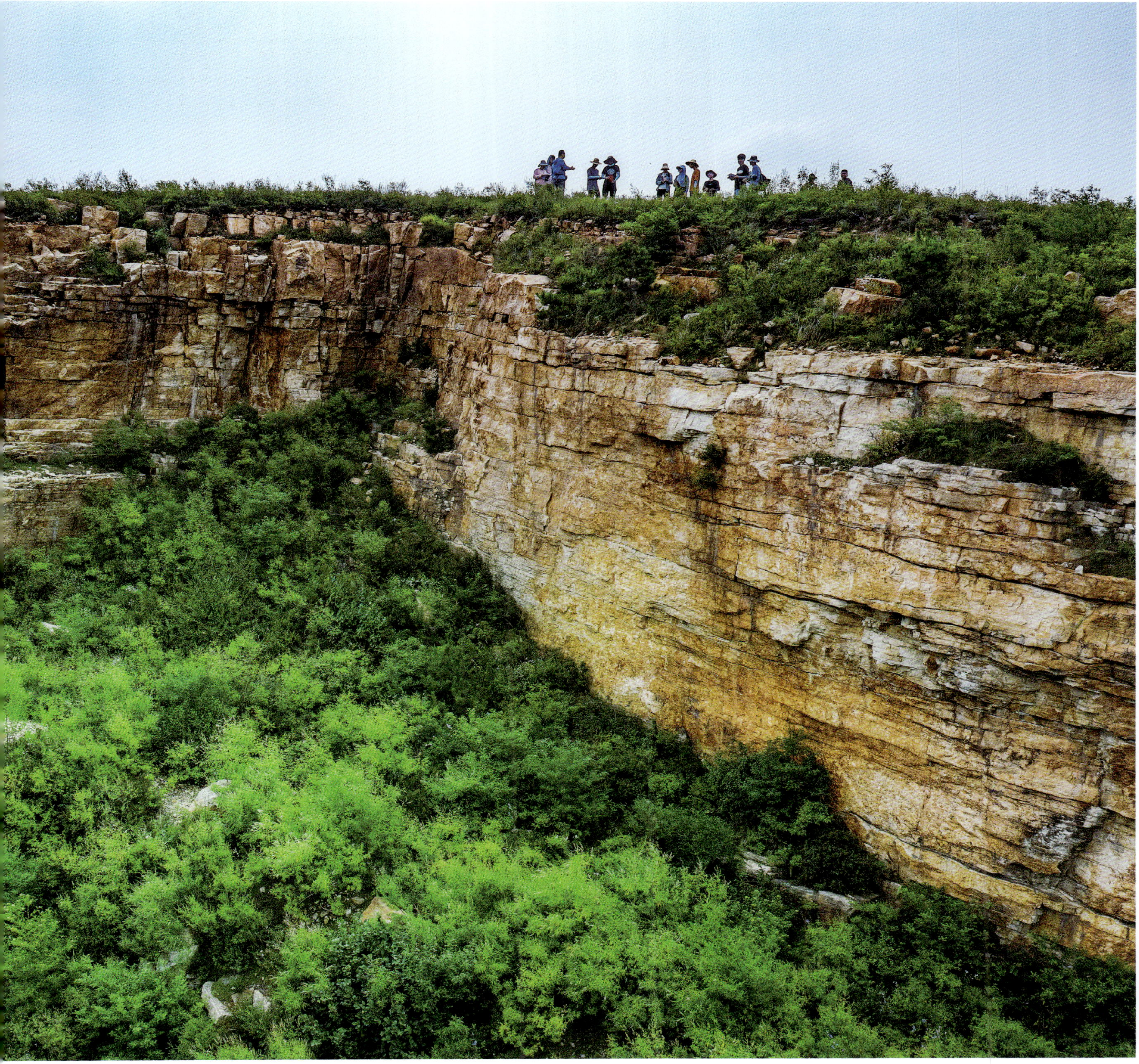

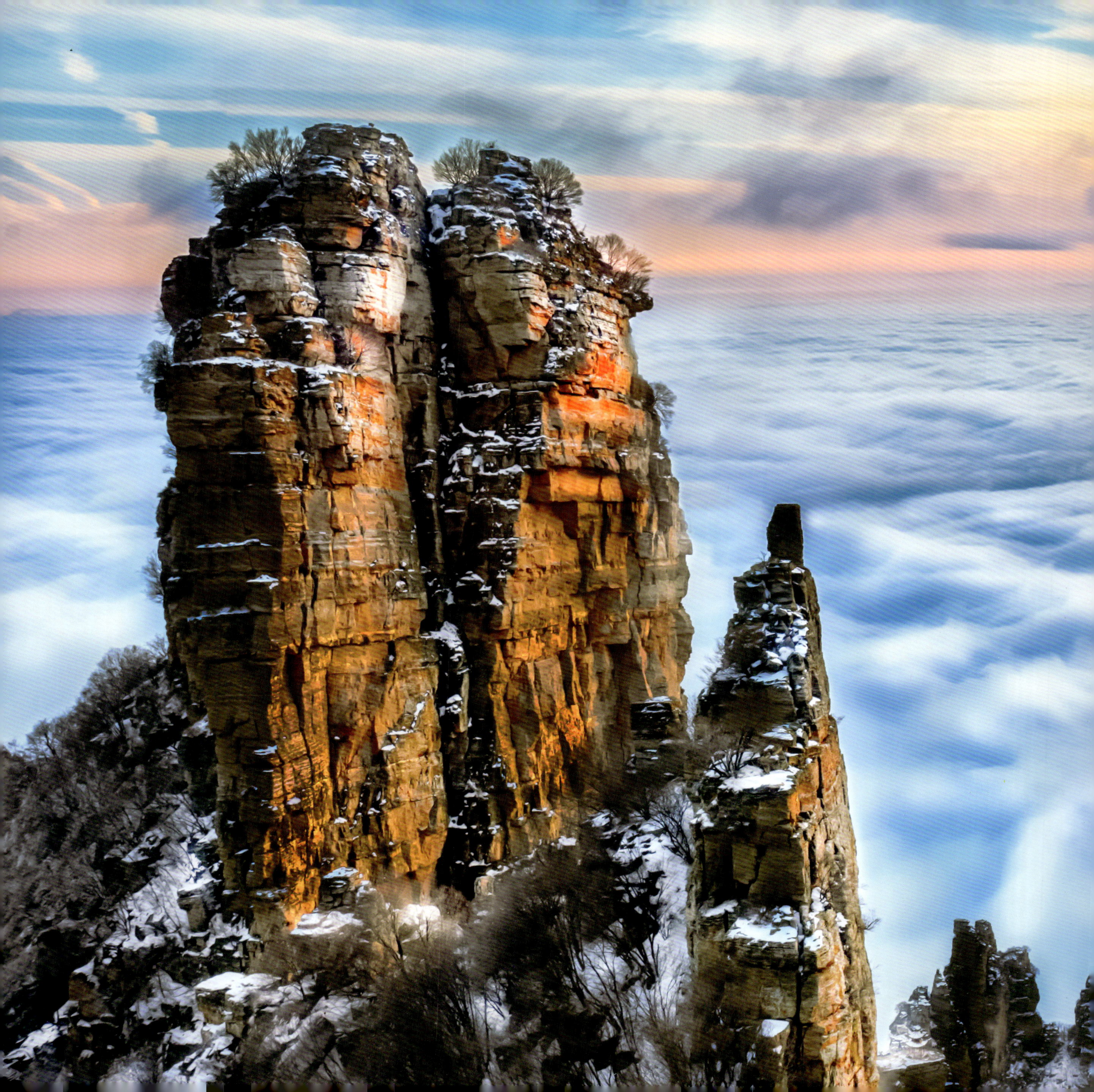

河北茅荆坝国家级自然保护区

　　河北茅荆坝国家级自然保护区位于承德市隆化县境内，其主要保护对象为森林生态系统及其生物多样性、珍稀濒危物种及其栖息地、自然生态环境和滦河上游水源地。该保护区属森林生态系统类型自然保护区。
《河北茅荆坝国家级自然保护区》，承德市隆化县，摄影：李安利

Maojingba National Nature Reserve, Hebei Province
Maojingba National Nature Reserve is located in Longhua County, Chengde City, and the main protection objects here are forest ecosystem and its biodiversity, rare and endangered species and their habitats, natural ecological environment, Luanhe River upstream water resource. It is a forest ecosystem type nature reserve.
Maojingba National Nature Reserve of Hebei Province, Longhua County, Chengde City
Photographed by Li Anli

国家公园资源

地质遗迹是大自然赐予人类不可再生的地质自然遗产，是人类认识地质现象、推测地质环境和演变条件的重要依据，是人类恢复地质历史的主要参数。

河北省地质遗迹资源丰富且远近闻名（见表11-1）。这里种类繁多的地质遗迹为旅游观光、科学研究和开发利用提供了丰富的天然资源。

河北省处在地质遗迹规划区中的华北、辽河平原、晋冀山地及辽东山东半岛区，这里集中了许多具有重要价值的地貌、地质剖面和古生物化石发掘地。经国务院批准，河北省建立了3处国家级地质遗迹自然保护区；经省政府批准，建立4处省级地质遗迹保护区；经自然资源部批准，建立9处国家级地质公园。

National Park Resources

Geological relics are the non-renewable geological natural heritage given by nature to human beings. It is an important basis for human beings to understand geological phenomena, speculate on geological environment and evolution conditions, and are the main parameters for human restoration of geological history.

Hebei Province is rich in geological relics resources and is well-knownfar and wide(Table 11-1). The wide variety of geological relics provide rich natural resources for tourism, scientific research, development and utilization.

Hebei Province is located in the geological relics planning area of North China, Liaohe River Plain, Shanxi-Hebei Mountains and Shandong Peninsula of Liaodong, where many important geomorphologies, geological profiles and paleontological fossil sources are concentrated. With the approval of the State Council, Hebei Province has established three national geological relics nature reserves; With the approval of the provincial government, 4 provincial-level geological relics protection areas have been established; With the approval of the Ministry of Natural Resources, 9 national geoparks have been established.

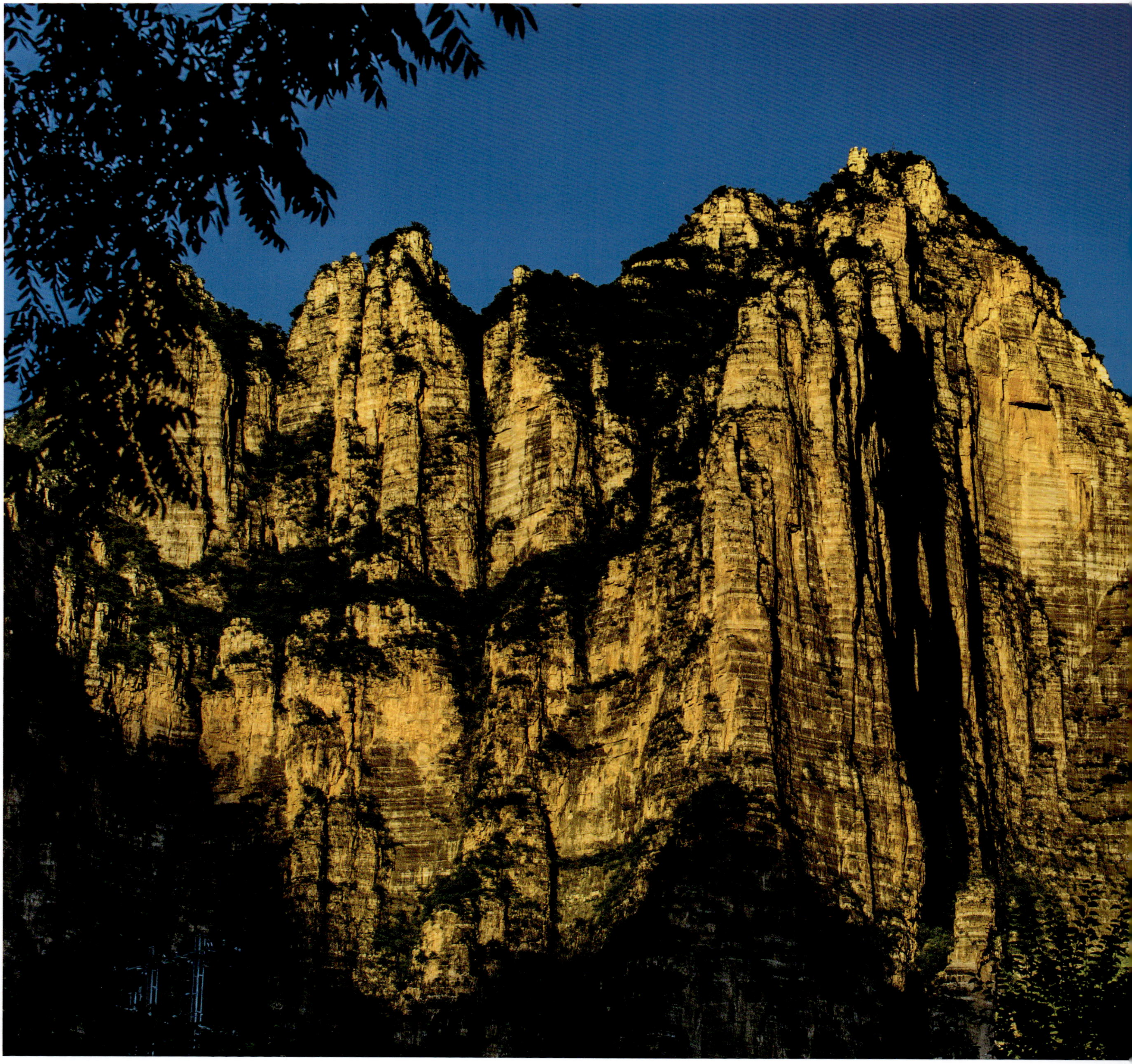

表 11-1 河北省地质公园
Table 11-1 Hebei Geopark

序号 Serial number	级别 Level	地质公园名称 Name of Geopark
1	国家级地质遗迹自然保护区 National Geological Relics Nature Reserve	柳江盆地地质遗迹国家级自然保护区 Liujiang Basin National Geological Heritage Nature Reserve
2		泥河湾国家级自然保护区 Nihewan National Nature Reserve
3		昌黎黄金海岸国家级自然保护区 Changli Gold Coast National Nature Reserve
4	省级地质遗迹保护区 Provincial Geological Heritage Reserve	海兴小山火山地质遗迹省级自然保护区 Haixing Xiaoshan Volcano Provincial Geological Relics Nature Reserve
5		丰宁四古生物化石省级自然保护区 Fengning Four Paleontological Fossil Provincial Nature Reserve
6		乐亭石臼坨诸岛省级海洋自然保护区 Leting Shijiutuo Islands Provincial Marine Nature Reserve
7		黄骅古贝壳堤省级海洋自然保护区 Huanghua Ancient Shell Provincial Marine Nature Reserve
8	国家级地质公园 National Geopark	河北野三坡世界地质公园 Hebei Yesanpo Global Geopark
9		河北武安国家地质公园 Hebei Wu'an National Geopark
10		河北赞皇嶂石岩国家地质公园 Hebei Zanhuangzhang Shiyan National Geopark

中国房山联合国教科文组织世界地质公园

中国房山联合国教科文组织世界地质公园位于北京市西南与河北省交界处，跨越北京市房山区，河北省涞水县、涞源县、易县两省市四区县。公园保存有距今 35 亿年以来完整的地层序列，系统地记录了地质历史变化，保存有多种典型的岩石类型。中国房山联合国教科文组织世界地质公园拥有举世无双的地质遗迹和文化遗产，是真正的"地质百科大全"。
《中国房山联合国教科文组织世界地质公园》，北京市西南与河北省交界处，摄影：曹树军

Fangshan UNESCO Global Geopark of China
Fangshan UNESCO Global Geopark is located on the border of Beijing southwest and Hebei Province. Across Fangshan District of Beijing; Laishui County, Laiyuan County and Yi County of Hebei. The park has preserved a complete stratigraphic sequence dating back 3.5 billion years, systematically recording geological historical changes, and preserving a variety of typical rock types. The China Fangshan UNESCO Global Geopark is a true "encyclopedia of geology" with unique geological relics and cultural heritage.
Fangshan UNESCO Global Geopark, Southwest of Beijing and Hebei Province
Photographed by Cao Shujun

河北秦皇岛柳江国家地质公园

河北秦皇岛柳江国家地质公园位于河北省秦皇岛市，南临渤海，北依燕山。2002年2月，国土资源部批准这里为第二批国家地质公园。该公园以古生物化石、地层剖面、岩溶地貌和花岗岩地貌为特色。这里地层剖面界限清楚，化石丰富，沉积构造发育充分。

《河北秦皇岛柳江国家地质公园》，秦皇岛市，摄影：崔重辉

Liujiang National Geopark, Qinhuangdao, Hebei Province

Liujiang National Geopark is located in Qinhuangdao City, Hebei Province, bordering the Bohai Sea in the south and Yanshan Mountain in the north. In February 2002, the Ministry of Land and Resources approved it as the second batch of national geoparks. It is characterized by paleontological fossils, stratigraphic sections, karst landforms and granite landforms. The boundaries of the stratigraphic section here are clear, fossils are abundant and sedimentary structures are developed.

Liujiang National Geopark, Qinhuangdao, Qinhuangdao City
Photographed by Cui Chonghui

河北阜平天生桥国家地质公园

　　河北阜平天生桥国家地质公园位于河北省阜平县城西约25千米处。2002年2月，国土资源部批准这里为第二批国家地质公园。这里以25亿年前的太古宇阜平群标准剖面和天生桥瀑布群等地质地貌景观为特色。

《河北阜平天生桥国家地质公园》，保定市阜平县，摄影：李玉亮

Tianshengqiao National Geopark, Fuping, Hebei Province

Tiansheng Bridge National Geopark is located in Fuping County, Hebei Province, about 25 kilometers west of the city. In February 2002, the Ministry of Land and Resources approved it as the second batch of national geoparks. It is characterized by the standard section of Archaic Realm fuping Group 2.5 billion years ago and Tiansheng Bridge Waterfall Group.

Tianshengqiao National Geopark, Fuping, Fuping County, Baoding City
Photographed by Li Yuliang

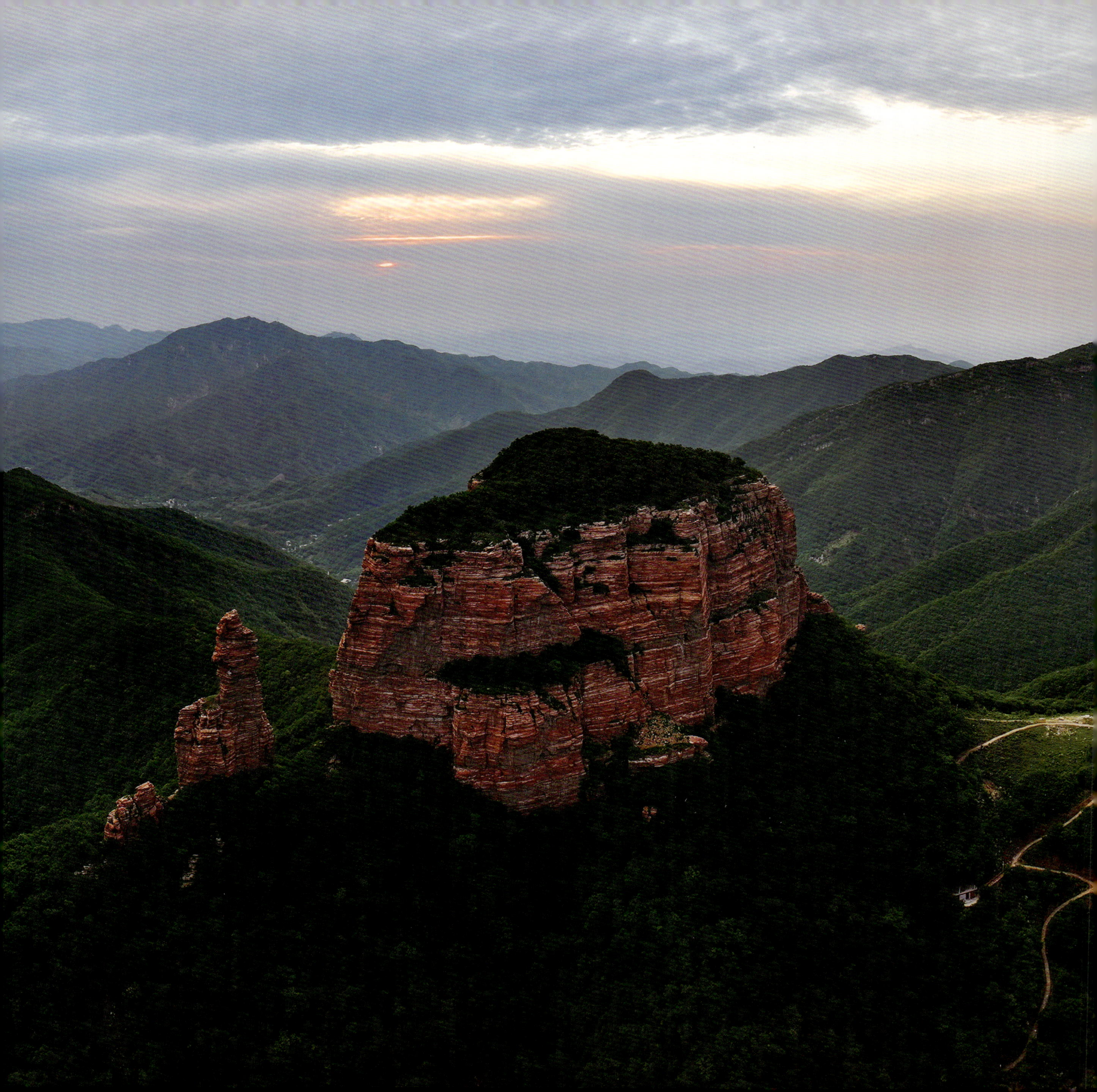

河北赞皇嶂石岩国家地质公园

河北赞皇嶂石岩国家地质公园位于河北省石家庄市赞皇县，地处太行山主脉中段槐河上游。2004年3月，国土资源部批准这里为第三批国家地质公园。公园内最为典型的是嶂石岩地貌和元古宇长城系砂岩中的层理与层面构造（交错层构造、波纹构造）。
《河北赞皇嶂石岩国家地质公园》，石家庄市赞皇县，摄影：冯建军

Zanhuang Zhangshiyan National Geopark, Hebei Province
Zanhuang Zhangshiyan National Geopark is located in Zhanhuang County, Shijiazhuang City, Hebei Province, and is situated in the middle of the main vein of Taihang Mountain and the upper reaches of Huaihe River. In March 2004, the Ministry of Land and Resources approved it as the third batch of national geoparks. The most typical features in the park are the Zhangshiyan landform and the stratification and bedding structure (cross-bedding structure, corrugated structure) in the Proterozoic Changcheng System sandstone.
Hebei Zanhuangzhang Shiyan National Geopark, Zanhuang County, Shijiazhuang City
Photographed by Feng Jianjun

■ 河北赞皇嶂石岩国家地质公园

《云漫仙人峰》，石家庄市赞皇县，摄影：程丙军

Zanhuang Zhangshiyan National Geopark, Hebei Province
Cloud over Fairy Peak, Zanhuang County, Shijiazhuang City
Photographed by Cheng Bingjun

河北赞皇嶂石岩国家地质公园

《嶂石岩地貌》，石家庄市赞皇县，摄影：丁建军

Zanhuang Zhangshiyan National Geopark, Hebei Province
Zhangshiyan Landform, Zanhuang County, Shijiazhuang City
Photographed by Ding Jianjun

河北临城国家地质公园

河北临城国家地质公园位于邢台市临城县中西部。这里以岩溶洞穴为主体，以稀有、奇特的天然岩溶洞穴和岩溶洞穴沉积景观遗迹为特色。

《河北临城国家地质公园》，邢台市临城县，摄影：李长春

Lincheng National Geopark, Hebei Province

Lincheng National Geopark is located in the Midwest of Lincheng County, Xingtai City, Hebei Province. With karst caves as the main part, there is characterized by the rare and strange natural karst caves andkarst cave sedimentary landscape remains.

Hebei Lincheng National Geopark, Lincheng County, Xingtai City
Photographed by Li Changchun

河北武安国家地质公园

　　河北武安国家地质公园位于邯郸市武安市西北 30 千米。这里以中古界长城系石英砂岩形成的水上红色峡谷峰林、宽谷峰林地貌景观为特色。

《河北武安国家地质公园》，武安市，摄影：宋现彬

Wu 'an National Geopark, Hebei Province

Wu 'an National Geopark is located in Handan City, 30 northwest kilometers near Wu 'an City. There is characterized by the red canyon on the water surface formed by the quartz sandstone of the Changcheng system in the Mesoproterozoic and the landform landscape of the wide valley peak forest.

Hebei Wu'an National Geopark, Wu'an City
Photographed by Song Xianbin

河北迁安迁西国家地质公园

迁西国家地质公园位于河北省唐山市迁西县。这里以古老的地质遗迹、奇特的地貌景观和悠久的历史文化而著称。
《河北迁安迁西国家地质公园》，唐山市迁西县，摄影：李彬

Qianxi National Geopark, Hebei Province

Qianxi National Geopark is located in Qianxi County, Tangshan City, Hebei Province. It is famous for archaic geological remains, strange landscape, long history and culture.
Hebei Qian'an Qianxi National Geopark, Qianxi County, Tangshan City
Photographed by Li Bin

河北兴隆国家地质公园

兴隆国家地质公园位于河北省兴隆县。这里是集燕山地质构造、岩溶洞穴、花岗岩景观、典型的地质构造剖面于一体的综合性地质公园。

《河北兴隆国家地质公园》，承德市兴隆县，摄影：衣志坚

Xinglong National Geopark, Hebei Province
Xinglong National Geopark is located in Xinglong County. This is a comprehensive geological park that integrates Yanshan geological structure, karst caves, granite landscape and typical geological structure sections.
Xinglong National Geopark, Hebei Province, Xinglong County, Chengde City
Photographed by Yi Zhijian

美景赏析

Appreciation of Beautiful Scenery

《龙凤呈祥》，唐山市南湖景区，摄影：孟东红

The dragon and phoenix are auspicious, Nanhu Scenic Area of Tangshan City
Photographed by Meng Donghong

雄安新区
Xiong'an New Area

《雄安新区千年秀林》，雄安新区千年秀林，摄影：胡忠

Millennium Show Forest in Xiong'an New Area, Xiong'an millennium beautiful forest
Photographed by Hu Zhong

《白洋淀燕南堤》，白洋淀燕南堤，摄影：胡忠

Yannan Dike in Baiyangdian, Baiyangdian Yannan Embankment
Photographed by Hu Zhong

《雄安千年秀林》，雄安雄安新区千年秀林，摄影：张学农

Xiong'an Millennium Show Forest, Xiong'an millennium beautiful forest
Photographed by Zhang Xuenong

《雄安郊野公园》，雄安雄安新郊野公园，摄影：张学农

Xiong'an Suburban Park, Xiong'an Country Park
Photographed by Zhang Xuenong

《雄安中央绿谷》，雄安新区中央绿谷，摄影：胡忠

Xiong'an Central Green Valley, Central Green Valley, Xiong'an New Area
Photographed by Hu Zhong

《雄安新区容东安置区悦容公园》，雄安新区悦容公园，摄影：胡忠

Yuerong Park in Rongdong Resettlement Area of Xiong'an New Area, Yuerong Park
Photographed by Hu Zhong

《雄安雪后夜景》，雄安新区悦容公园夜景，摄影：张学农

Night Scenery of Xiong'an After Snow, Night view of Yuerong Park
Photographed by Zhang Xuenong

《雄安商服会议中心》，雄安新区服务中心，摄影：张学农

Xiong'an Business Service Conference Center, Xiong'an Service Center
Photographed by Zhang Xuenong

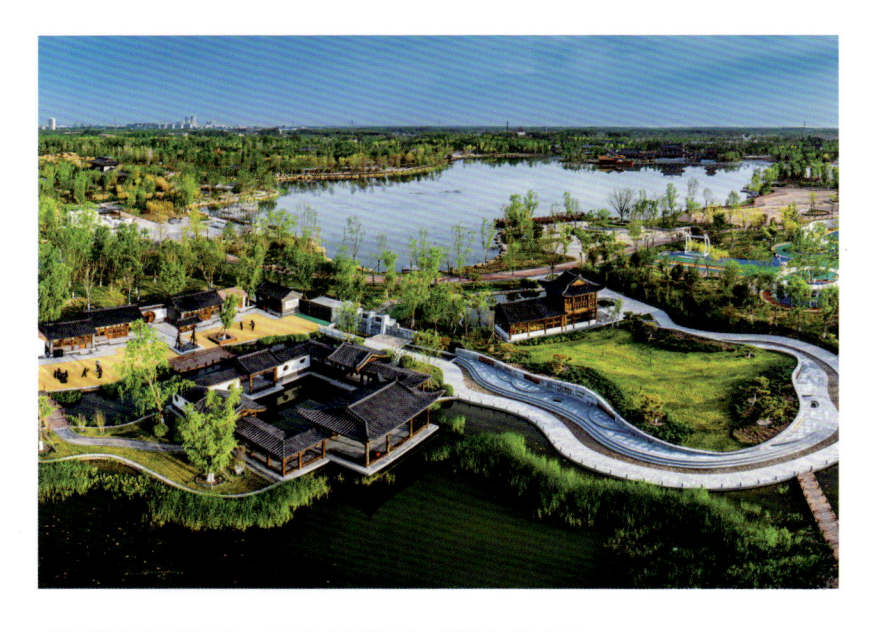

《沧州园博园风光》，沧州市园博园，摄影：陈秀峰

Scenery of Cangzhou Garden Expo Park, Garden Expo Park, Cangzhou City
Photographed by Chen Xiufeng

《国际庄》，石家庄市解放广场，摄影：赵宇涵

International Village, Liberation Square, Shijiazhuang City
Photography by Zhao Yuhan

《保定新地标》，保定市拾光公园，摄影：梁凤华

New Landmark of Baoding, Shiguang Park, Baoding City
Photographed by Liang Fenghua

《保定古莲花池》，保定市莲池区，摄影：李康

Baoding Ancient Lotus Pond, Lianchi District, Baoding City
Photographed by Li Kang

《唐山市迁西县潘家口水下长城》，唐山市迁西县潘家口水库潘家口水上长城，摄影：王爱军

Underwater Great Wall at Panjiakou, Qianxi County, Tangshan City
Panjiakou Reservoir, Panjiakou Water Great Wall, Qianxi County, Tangshan City
Photographed by Wang Aijun

《仙境茶山村》，张家口市蔚县茶山，摄影：马宝玺

Fairyland Tea Mountain Village, Tea Mountain, Wei County, Zhangjiakou City
Photographed by Ma Baoxi

《希望的田野》，廊坊市文安县，摄影：张永康

Field of Hope, Wen'an County, Langfang City
Photographed by Zhang Yongkang

《运河两岸大丰收》,衡水市景县留智庙镇,摄影:郑文景

Liuzhimiao Town, Jing County, Liuzhimiao Town, Jing County, Hengshui City
Photographed by Zheng Wenjing

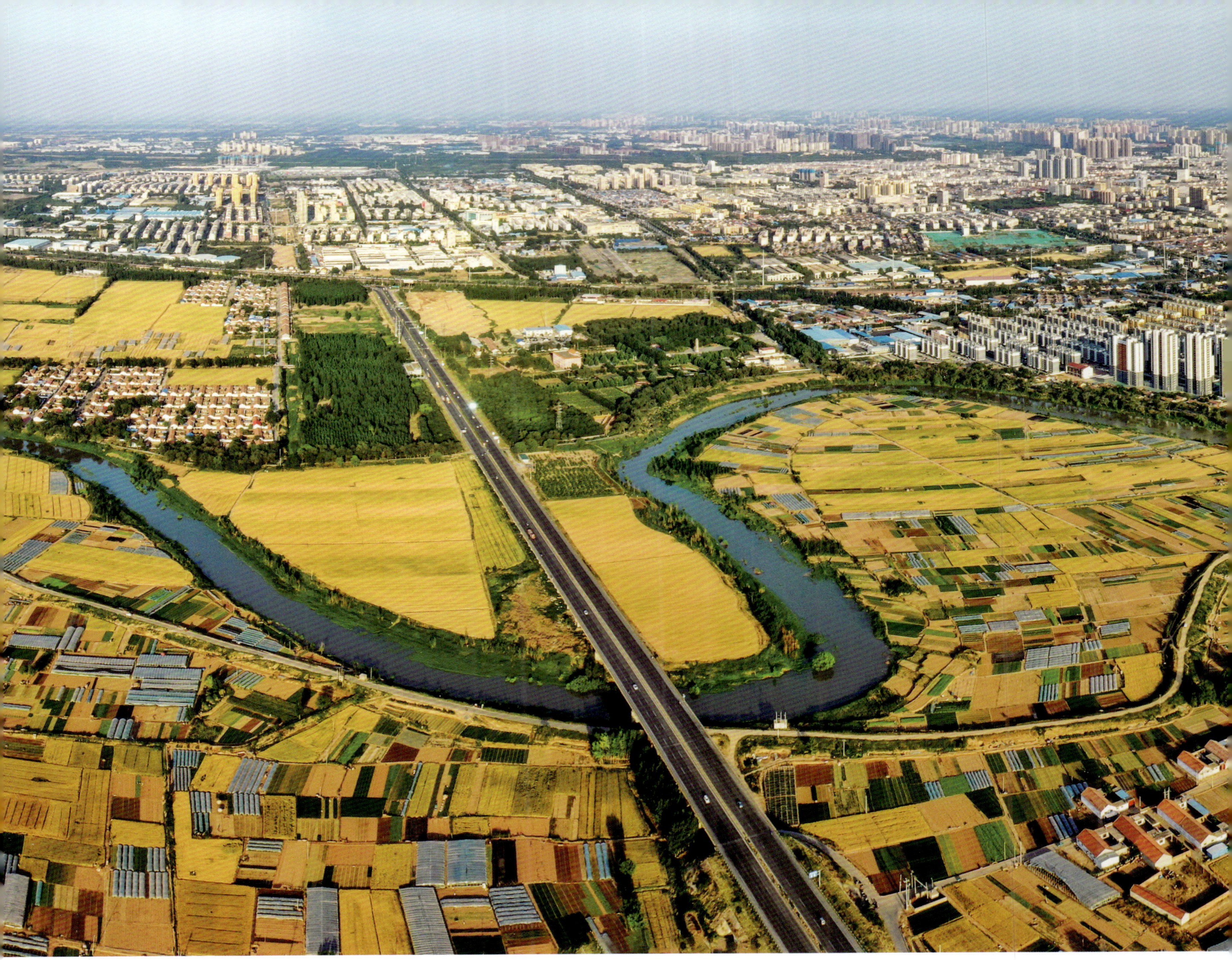

《长城》，金山岭长城，摄影：李秀清

The Great Wall, Jinshanling Great Wall
Photographed by Li Xiuqing

《赤壁丹崖九女峰》，石家庄市赞皇县，摄影：霍红军

Nine Maidens Peak at Red Cliffs, Zanhuang County, Shijiazhuang City
Photographed by Huo Hongjun

《火山岩上的长城》，秦皇岛市海港区板厂峪长城，摄影：王守民

The Great Wall on Volcanic Rocks, The Great Wall of Banchangyu, Haigang District, Qinhuangdao City
Photographed by Wang Shoumin

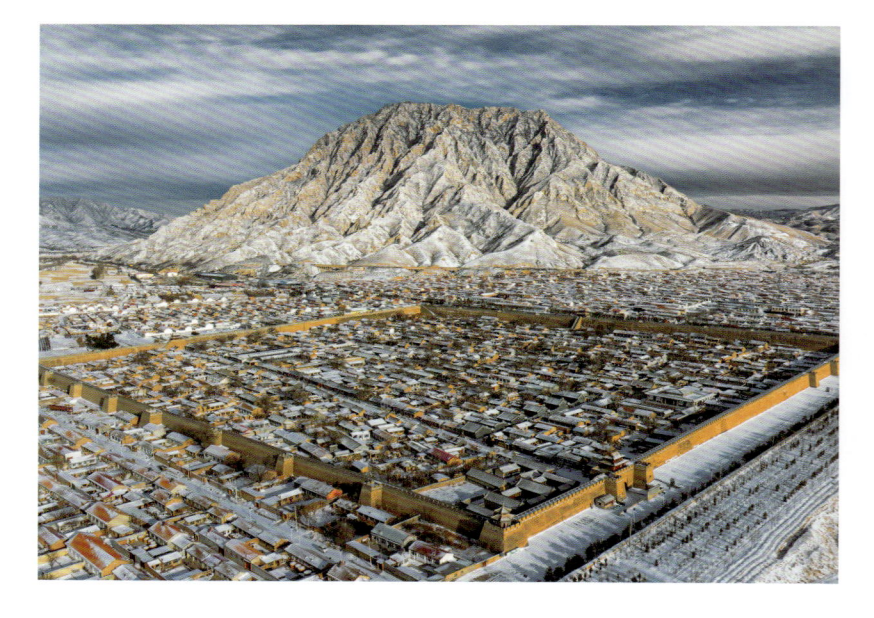

《鸡鸣山下鸡鸣驿》，张家口市怀来县鸡鸣驿，摄影：董桂军

Jiming Posthouse at the Foot of Jiming Mountain, Zhangjiakou City, Huailai County, Jimingyi
Photographed by Dong Guijun

《夕阳海神庙》，秦皇岛市山海关老龙头，摄影：刘丽

Sunset at the Sea God Temple, The old leader of Shanhaiguan in Qinhuangdao City
Photographed by Liu Li

《云雾缥缈》，清东陵，慈禧和慈安陵，摄影：孟宪华

Misty Clouds - Qing Dongling, Cixi and Ci'an Mausoleum
Photographed by Meng Xianhua

《河北宣化区大峡谷》，张家口市宣化区大峡谷像光洞，摄影：罗文琛

Grand Canyon in Xuanhua District, Hebei Province, The Grand Canyon in Xuanhua District, Zhangjiakou City is like a light cave
Photographed by Luo Wenchen

《吕家古村柿柿红》，石家庄市井陉县吕家古村，摄影：郭西力

Holy Land of the War of Resistance, Lujia Ancient Village, Jingcheng County, Shijiazhuang City
Photographed by Guo Xili

《夜幕》，张家口市崇礼区太舞度假小镇，摄影：李登云
Nightfall, Taiwu Resort Town, Chongli District, Zhangjiakou City
Photographed by Li Dengyun

《冬览碧螺塔》，秦皇岛市北戴河碧螺塔公园，摄影：蔡鹤逊
Winter Glimpse of Biluo Pagoda, Biluota Park, Beidaihe, Qinhuangdao City
Photographed by Cai Hexun

《骆驼湾夜色》，保定市阜平县骆驼湾村，摄影：李玉亮
Night View of Luotuo Bay, Camel Bay Village, Fuping County, Baoding City
Photographed by Li Yuliang

《崆山白云洞》，邢台市临城县，摄影：陈志杰
Kong Baiyun Mountain Cave, Lincheng County, Xingtai City
Photographed by Chen Zhijie

《唐山新高铁站》，唐山市，摄影：朱元杰

Tangshan new high-speed railway station, Tangshan City
Photographed by Zhu Yuanjie

《张家口万家灯火》，张家口市，摄影：李晓宁

Zhangjiakou's Thousands of Lights, Zhangjiakou City
Photographed by Li Xiaoning

《魅力邯郸》，邯郸市美乐城龙湖公园，摄影：邵峰

Charming Handan, Longhu Park, Meile City, Handan City
Photographed by Shao Feng

《蔚蓝海岸》，秦皇岛市蔚蓝海岸，摄影：李冰

Blue Coast, Qinhuangdao City Cote d'Azur
Photographed by Li Bing

225

《假日北戴河》，秦皇岛市北戴河老虎石公园，摄影：崔重辉

Holiday in Beidaihe, Tiger Stone Park, Beidaihe, Qinhuangdao City
Photographed by Cui Zhonghui

《辛集润泽湖公园》，辛集市辛集润泽湖公园，摄影：种勇

Xinji Runze Lake Park, Xinji Runze Lake Park, Xinji City
Photographed by Zhong Yong

《遗鸥岛》，张家口市康保县康巴诺尔遗鸥岛，摄影：赵永春

Island of Lost Gulls, Kangbao County, Zhangjiakou City, Kangbanuoer Relict Gull Island
Photographed by Zhao Yongchun

《生态白洋淀》，雄安新区，摄影：贺友顺

Ecological Baiyangdian, Xiong'an New Area
Photographed by He Youshun

《避暑山庄》，承德市避暑山庄，摄影：于磊

Mountain Resort, Chengde City Summer Resort
Photographed by Yu Lei

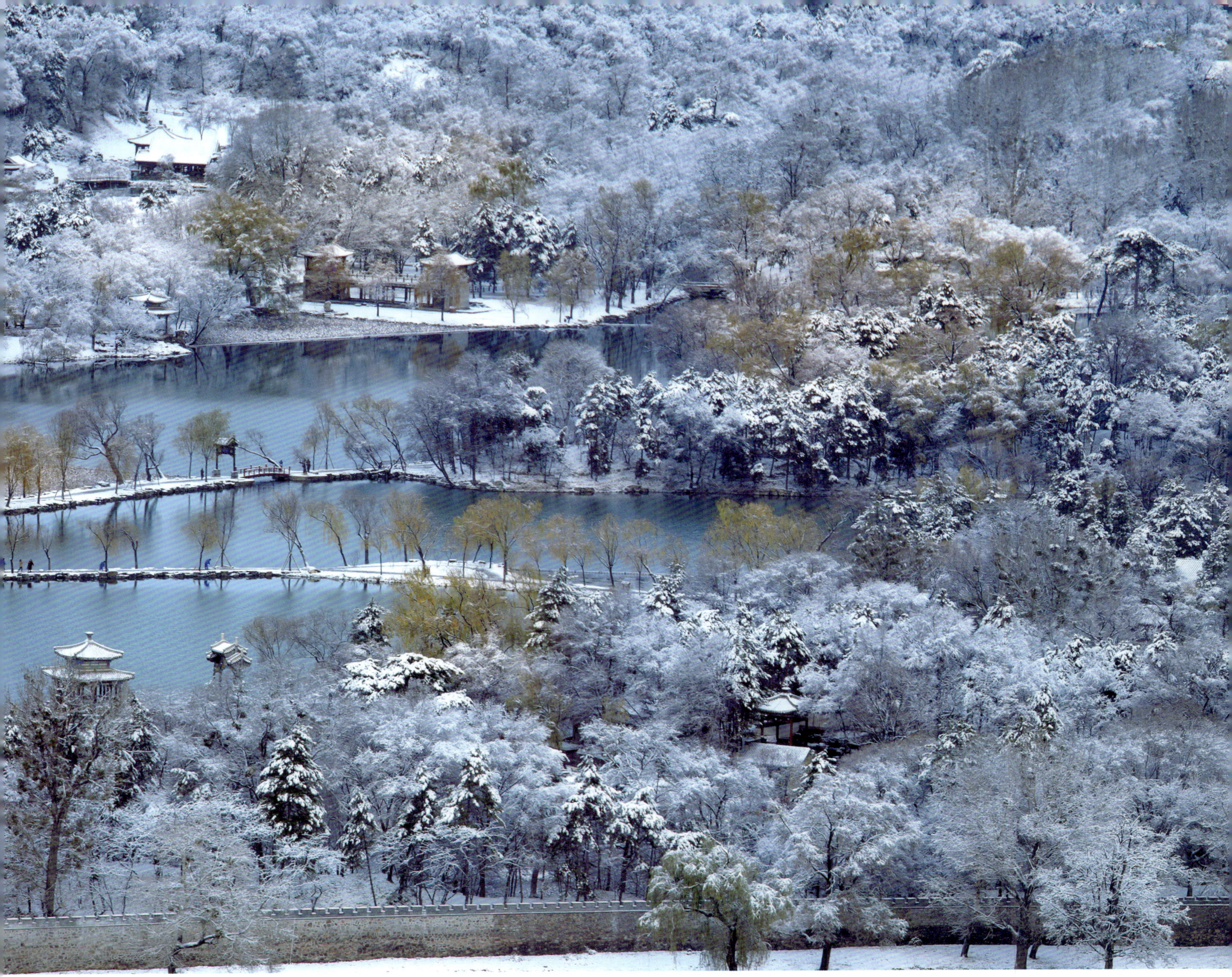

后 记

《河北自然资源图鉴》终于与读者见面了。这是河北自然资源厅宣传中心宣传项目中的又一力作。

在《河北自然资源图鉴》的编辑过程中，我们得到了河北省摄影家协会和全国各地摄影师的积极响应与支持。在众多精彩的影像作品中，我们选出 205 幅作品编辑成册。

本书围绕河北省自然资源管理的新职责、新使命和新任务，坚持"山水林田湖草生命共同体"理念，按照既全面系统，又突出重点的设计原则，收集整理、拍摄提炼可公开的河北自然资源调查监测数据和相关地图成果，分"国土空间规划""土地资源""矿产资源""森林资源""草原资源""湿地资源""水资源""海洋资源""野生动物资源""自然保护区资源""国家公园资源"和"美景赏析"12 个专题，全方位展现河北省自然资源空间分布、开发利用、保护监管以及与之相关的丰富信息。

《河北自然资源图鉴》在全体编纂人员的共同努力下完成，得到了众多专家的悉心指导和帮助，由于时间、篇幅等客观因素的限制，未能做到面面俱到，不当之处在所难免，恳请读者批评指正。

在此，以期通过本书的出版让社会各界进一步关注河北、关注河北自然资源工作，希望广大读者以实际行动共同珍惜自然资源、合理利用自然资源、保护自然资源。山积而高，泽积而长。

2024 年是中华人民共和国成立 75 周年，谨以此书向祖国 75 华诞献礼！

Postscript

The *Hebei Natural Resources Illustrations Handbook* is finally met with readers, which is another outstanding work of the Publicity center Department of Natural Resources of Hebei Province.

In the editing process of *Hebei Natural Resources Illustrations Handbook*, we have received positive response and support from Hebei Photographers Association and photographers all over the country. Among the many wonderful photographic works, we have selected 205 works and compiled them into a book.

The book focuses on the new responsibilities, new missions and new tasks of natural resource management in Hebei Province, adhere to the concept of "Mountains, rivers, forests, farmlands, lakes and grasslands are a community of life", according to the design principles of considering comprehensive system and highlighting key points, collected and sorted photograph and refine publicly available survey and monitoring data and related map results of Hebei natural resources, which including 12 parts: "Territorial spatial planning""land resources""mineral resources""forest resources""grassland resources""wetland resources""Marine resources""wildlife resources""Natural reserve resources""National park resources""water resources" and "Appreciate the beautiful scenery", comprehensively presents the spatial distribution of natural resources, development and utilization, protection and supervision of natural resources in Hebei Province, as well as rich information related to it.

Hebei Natural Resources Illustrations Handbook was completed under the joint efforts of all team members, and received the careful guidance and help of many experts. Due to the limitations of objective factors such as time and length, it inevitably has omissions and shortcomings, readers are invited to criticize.

Here, through the publication of this book, we hope that the society pay more attention to Hebei, pay more attention to Hebei natural resources work, hope that the readers will take practical actions to cherish natural resources, rationally use natural resources, and protect natural resources.

The year 2024 marks the 75th anniversary of the founding of the People's Republic of China, this book is a tribute to our motherland!

图书在版编目（CIP）数据

河北自然资源图鉴：汉、英 / 章洪涛主编.
北京：中国摄影出版传媒有限责任公司，2024.12.
ISBN 978-7-5179-1502-7

Ⅰ．P966.222-64

中国国家版本馆 CIP 数据核字第 20251YG141 号

书　　名：	河北自然资源图鉴
主　　编：	章洪涛
总 策 划：	杨淑梅、任白羽
策划编辑：	王彪
责任编辑：	宋蕊
美术编辑：	任白羽
装帧设计：	王剑芬、任白羽
翻　　译：	张阳、孙艺芮
封面摄影：	张潇飞

出　　版：中国摄影出版传媒有限责任公司（中国摄影出版社）
　　地　　址：北京东城区东四十二条 48 号　邮编：100007
　　发行部：010-55136125 65280977
　　网　　址：www.cpph.com
　　邮　　箱：distribution@cpph.com

印　　刷：石家庄真彩印业有限公司
纸张规格： 787mm×1092mm
开　　本：12 开
印　　张：19.5
版　　次：2024 年 12 月第 1 版
印　　次：2024 年 12 月第 1 次印刷
印　　数：1—1000 册
ISBN 978-7-5179-1502-7
定　　价：268.00 元

版权所有　　侵权必究